MOON LORE

Myths, Worship, and Superstitions

By

TIMOTHY HARLEY

First published in 1885

Published by Wyrd Books,
an imprint of Read & Co.

Moon Lore first published in 1885
This edition published by Read & Co. in 2024

A catalogue record for this book is available from the British Library.

ISBN: 9781528724111

Read & Co. is part of Read Books Ltd.
For more information visit www.wyrdbooks.co.uk

"And when the clear moon, with its soothing influences, rises full in my view,—from the wall-like rocks, out of the damp underwood, the silvery forms of past ages hover up to me, and soften the austere pleasure of contemplation."

Goethe's *"Faust." Hayward's Translation,*
London, 1855, *p.* 100.

Contents

Moon Lore

The Literature of Light
1887

The Rev. Timothy Harley, F.R.A.S., who writes *Moon-Lore* is endowed with a capacity for collecting, assimilating, and making the most of out-of-the-way knowledge. It is not quite in jocularity that he styles this volume "a contribution to light literature, and to the literature of light."

Some of Mr. Harley's jokes smell of the lamp, "others have a look" of the nursery, and with several most readers would in any case have gladly dispensed. Mr. Harley might also have spared his readers a good deal of half-comic, half-serious hairsplitting—useful, perhaps, as a contribution to what Mrs. Carlyle termed "coterie-talk," but not otherwise—like this:—"Religion is a science, and science is a religion; but they are not identical. Philosophy ought to be pions, and piety ought to be philosophical; but philosophy and piety are two quantities and qualities that may dwell apart, though happily they may also be found in one nature."

But when the reader has become accustomed to, or learned to wink at, Mr. Harley's eccentricities in style, he will find this book very entertaining as a collection of anecdotes, and other curious information, on moon-spots (including the Man in the Moon, the Woman in the Moon, the Hare in the Moon, and the Toad in the Moon), moon-worship, moon-superstitions, and moon-inhabitation.

Mr. Harley's volume is, indeed, rather one to be dipped into at odd moments, like Barton's "Anatomy of Melancholy," or Isaac Disraeli's a books.

Thus, folks who read while running may think it odd that the moon is mostly a male and not a female deity. Yet such is the case. The Arabs, the Romans, and the Germans represent the moon as a male.

"In Slavonic," Sir George Cox says, "as in the Tectonic mythology, the moon is male." Also "among the Mbocobis of South America, the moon is a man and the sun his wife;" while "the Ahts of North America take the same view; and we know that in Sanscrit and in Hebrew the word for moon is masculine." This is but a specimen of Mr. Harley's book, the delightful character of which is enhanced by its curious little illustrations.

The Spectator, 26 March 1887

To the Moon

Composed by the Sea-Side—On the Coast of Cumberland

Wanderer! that stoop'st so low, and com'st so near
To human life's unsettled atmosphere;
Who lov'st with Night and Silence to partake,
So might it seem, the cares of them that wake;
And, through the cottage-lattice softly peeping,
Dost shield from harm the humblest of the sleeping;
What pleasure once encompassed those sweet names
Which yet in thy behalf the Poet claims,
An idolizing dreamer as of yore!—
I slight them all; and, on this sea-beat shore
Sole-sitting, only can to thoughts attend
That bid me hail thee as the Sailor's Friend;
So call thee for heaven's grace through thee made known
By confidence supplied and mercy shown,
When not a twinkling star or beacon's light
Abates the perils of a stormy night;
And for less obvious benefits, that find
Their way, with thy pure help, to heart and mind;
Both for the adventurer starting in life's prime;
And veteran ranging round from clime to clime,
Long-baffled hope's slow fever in his veins,
And wounds and weakness oft his labour's sole remains.

The aspiring Mountains and the winding Streams,
Empress of Night! are gladdened by thy beams;
A look of thine the wilderness pervades,
And penetrates the forest's inmost shades;
Thou, chequering peaceably the minster's gloom,
Guid'st the pale Mourner to the lost one's tomb;
Canst reach the Prisoner—to his grated cell

Welcome, though silent and intangible!—
And lives there one, of all that come and go
On the great waters toiling to and fro,
One, who has watched thee at some quiet hour
Enthroned aloft in undisputed power,
Or crossed by vapoury streaks and clouds that move
Catching the lustre they in part reprove—
Nor sometimes felt a fitness in thy sway
To call up thoughts that shun the glare of day,
And make the serious happier than the gay?

Yes, lovely Moon! if thou so mildly bright
Dost rouse, yet surely in thy own despite,
To fiercer mood the phrenzy-stricken brain,
Let me a compensating faith maintain;
That there's a sensitive, a tender, part
Which thou canst touch in every human heart,
For healing and composure.—But, as least
And mightiest billows ever have confessed
Thy domination; as the whole vast Sea
Feels through her lowest depths thy sovereignty;
So shines that countenance with especial grace
On them who urge the keel her *plains* to trace
Furrowing its way right onward. The most rude,
Cut off from home and country, may have stood—
Even till long gazing hath bedimmed his eye,
Or the mute rapture ended in a sigh—
Touched by accordance of thy placid cheer,
With some internal lights to memory dear,
Or fancies stealing forth to soothe the breast
Tired with its daily share of earth's unrest,—
Gentle awakenings, visitations meek;
A kindly influence whereof few will speak,
Though it can wet with tears the hardiest cheek.

And when thy beauty in the shadowy cave

Is hidden, buried in its monthly grave;
Then, while the Sailor, 'mid an open sea
Swept by a favouring wind that leaves thought free,
Paces the deck—no star perhaps in sight,
And nothing save the moving ship's own light
To cheer the long dark hours of vacant night—
Oft with his musings does thy image blend,
In his mind's eye thy crescent horns ascend,
And thou art still, O Moon, that Sailor's Friend!

—William Wordsworth, 1835

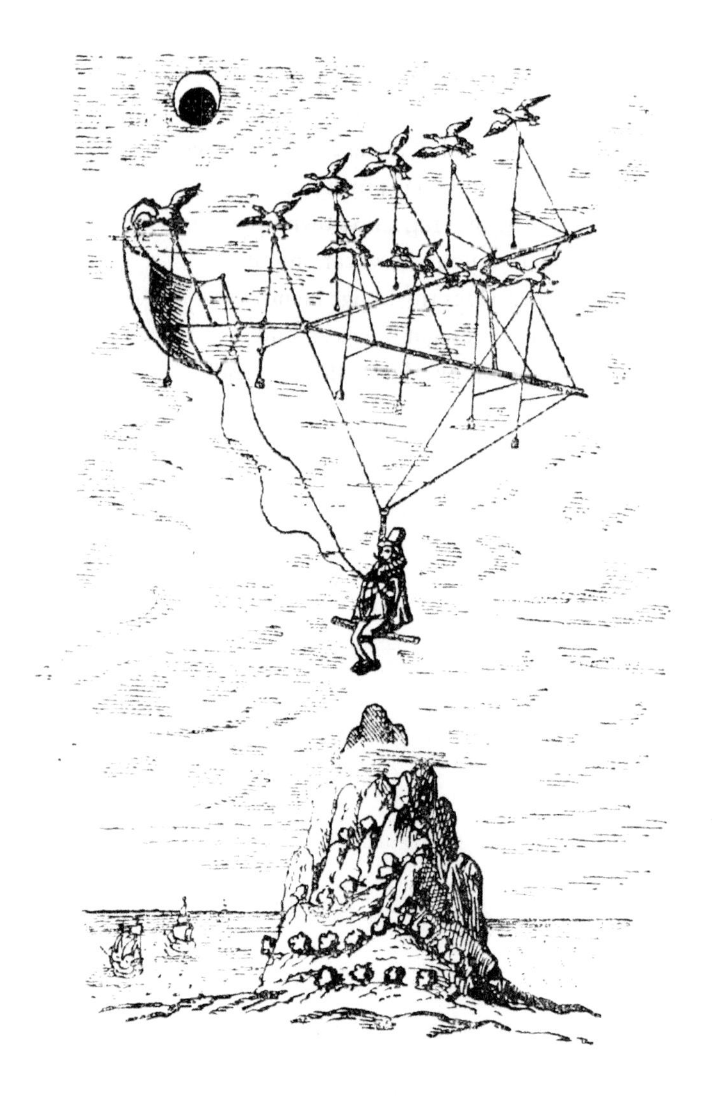

Voyaging to the Moon
From Domingo Gonsales (A.D. 1638)

MOON LORE

By Timothy Harley

"I beheld the moon walking in brightness."

—*Job* xxxi. 26.

"The moon and the stars, which Thou hast ordained."

—*Psalm* viii. 3.

"Who is she that looketh forth, fair as the moon?"

—*Solomon's Song* vi. 10.

"The precious things put forth by the moon."

—*Deuteronomy* xxxiii. 14.

"Soon as the evening shades prevail,
The moon takes up the wondrous tale."

—Addison's *Ode.*

"In fall-orbed glory, yonder moon Divine
Rolls through the dark-blue depths."

—Southey's T*halaba.*

"Queen of the silver bow! by thy pale beam,
  Alone and pensive, I delight to stray,
And watch thy shadow trembling in the stream,
  Or mark the floating clouds that cross thy way;
And while I gaze, thy mild and placid light
  Sheds a soft calm upon my troubled breast:
And oft I think-fair planet of the night—
  That in thy orb the wretched may have rest;
The sufferers of the earth perhaps may go—
  Released by death-to thy benignant sphere;
And the sad children of despair and woe
  Forget in thee their cup of sorrow here.
Oh that I soon may reach thy world serene,
  Poor wearied pilgrim in this toiling scene!"

—Charlotte Smith.

Preface

This work is a contribution to light literature, and to the literature of light. Though a monograph, it is also a medley.

The first part is mythological and mirthsome. It is the original nucleus around which the other parts have gathered. Some years since, the writer was led to investigate the world-wide myth of the Man in the Moon, in its legendary and ludicrous aspects; and one study being a stepping-stone to another, the ball was enlarged as it rolled.

The second part, dealing with moon-worship, is designed to show that anthropomorphism and sexuality have been the principal factors in that idolatry which in all ages has paid homage to the hosts of heaven, as *heaved* above the aspiring worshipper. Man adores what he regards as higher than he. And if the moon is supposed to affect his tides, that body becomes his water-god.

The third part treats of lunar superstitions, many of which yet live in the vagaries which sour and shade our modern sweetness and light.

The fourth and final part is a literary essay on lunar inhabitation, presenting *in nuce* the present state of the enigma of "the plurality of worlds."

Of the imperfections of his production the author is partly conscious. Not *wholly* so; for others see us often more advantageously than we see ourselves. But a hope is cherished that this work—a compendium of lunar literature in its least scientific branches—may win a welcome which shall constitute the worker's richest reward. To the innumerable writers who are quoted, the indebtedness felt is inexpressible.

MOON SPOTS

Introduction

With the invention of the telescope came an epoch in human history. To Hans Lippershey, a Dutch optician, is accorded the honour of having constructed the first astronomical telescope, which he made so early as the 2nd of October, 1608. Galileo, hearing of this new wonder, set to work, and produced and improved instrument, which he carried in triumph to Venice, where it occasioned the intensest delight. Sir David Brewster tells us that "the interest which the exhibition of the telescope excited at Venice did not soon subside: Sirturi describes it as amounting to frenzy. When he himself had succeeded in making one of these instruments, he ascended the tower of St. Mark, where he might use it without molestation. He was recognised, however, by a crowd in the street, and such was the eagerness of their curiosity, that they took possession of the wondrous tube, and detained the impatient philosopher for several hours till they had successively witnessed its effects."[1] it was in May, 1609, that Galileo turned his telescope on the moon. "The first observations of Galileo," says Flammarion, "did not make less noise than the discovery of America; many saw in them another discovery of a new world much more interesting than America, as it was beyond the earth. It is one of the most curious episodes of history, that of the prodigious excitement which was caused by the unveiling of the world of the moon."[2] Nor are we astonished at their astonishment when they beheld mountains which have since been found to be from 15,000 to 26,000 feet in height—highlands of the moon indeed—far higher in proportion to the moon's diameter than any elevations on the earth; when they saw the surface of the satellite scooped out into deep valleys, or spread over with vast walled plains from 130 to 140 miles across. No wonder that the followers of Aristotle resented the explosion of their preconceived beliefs; for their master had taught that the moon was perfectly spherical and smooth, and that the spots were merely reflections of our own mountains. Other ancient philosophers had said that these patches were shadows of opaque bodies floating between the sun and the moon. But to the credit of Democritus be it remembered that he propounded the opinion that the spots were diversities or inequalities upon the lunar surface; and thus anticipated by twenty centuries the disclosures of the telescope. The invention of this invaluable appliance we have regarded as marking a great modern epoch; and what is usually written on the moon is mainly a summary of results obtained through telescopic observation, aided by other apparatus, and conducted by learned men. We now purpose to go back to the ages when there were neither reflectors nor refractors in existence; and to travel beyond the bounds of ascertained fact into the regions of fiction, where abide the shades of superstition and the dreamy forms of myth. Having promised a contribution to light literature, we shall give to fancy a free rein, and levy taxes upon poets and story-tellers, wits and humorists wherever they may be of service. Much will have to be said, in the first place, of the man in the moon, whom we must view as he has been manifested in the mask of mirth, and also in the mirror of mythology. Then we shall present the woman in the moon, who is less known than the immortal man. Next a hare will be started; afterwards a frog, and other objects; and when we reach the end of our excursion, if we mistake not, it will be confessed that the moon has created more merriment, more marvel, and more mystery, than all of the other orbs taken together.

But before we forget the fair moon in the society of its famous man, let us soothe our spirits in sweet oblivion of discussions and dissertations, while we survey its argentine glories with poetic rapture. Like Shelley, we are all in love with

"That orbèd maiden, with white fire laden,
Whom mortals call the moon." (*The Cloud.*)

Our little loves, who take the lowest seats in the domestic synagogue, if they cannot have the moon by crying for it, will rush out, when they ought to be in bed, and chant,

"Boys and girls come out to play,
The moon doth shine as bright as day."

The young ladies of the family, without a tincture of affectation, will languish as they gaze on the lovely Luna. Not, as a grumpy, grisly old bear of a bachelor once said, "Because there's a man in it!" No; the precious pets are fond of moonlight rather because they are the daughters of Eve. They are in sympathy with all that is bright and beautiful in the heavens above, and in the earth beneath; and it has even been suspected that the only reason why they ever assume that invisible round-about called crinoline is that, like the moon, they may move in a circle. Our greatest men, likewise, are susceptible to Luna's blandishments. In proof of this we may produce a story told by Mark Lemon, at one time the able editor of Punch. By the way, an irrepressible propensity to play upon words has reminded some one that punch is always improved by the essence of lemon. But this we leave to the bibulous, and go on with the story. Lord Brougham, speaking of the salary attached to a new judgeship, said it was all moonshine. Lord Lyndhurst, in his dry and waggish way, remarked, "May be so, my Lord Harry; but I have a strong notion that, moonshine though it be, you would like to see the *first quarter* of it."[3] That Hibernian was a discriminating admirer of the moon who said that the sun was a coward, because he always went away as soon as it began to grow dark, and never came back till it was light again; while the blessed moon stayed with us through the forsaken night. And now, feeling refreshed with these exhilarating meditations, we, for awhile, leave this lovable orb to those astronomical stars who have studied the heavens from their earliest history; and hasten to make ourselves acquainted with the proper study of mankind, the ludicrous and legendary lunar man.

The Man in the Moon

We must not be misunderstood. By the man in the moon we do not mean any public tavern, or gin-palace, displaying that singular sign. The last inn of that name known to us in London stands in a narrow passage of that fashionable promenade called Regent Street, close to Piccadilly. Nor do we intend by the man in the moon the silvery individual who pays the election expenses, so long as the elector votes his ticket. Neither do we mean the mooney, or mad fellow who is too fond of the cup which cheers and then inebriates; nor even one who goes mooning round the world without a plan or purpose. No; if we are not too scientific, we are too straightforward to be allured by any such false lights as these. By the man in the moon we mean none other than that illustrious personage, whose shining countenance may be beheld many a night, clouds and fogs permitting, beaming good-naturedly on the dark earth, and singing, in the language of a lyric bard,

"The moon is out to-night, love,
Meet me with a smile."

But some sceptic may assail us with a note of interrogation, saying, "Is there a man in the moon?" "Why, of course, there is!" Those who have misgivings should ask a sailor; he knows, for the punsters assure us that he has been to *sea*. Or let them ask any *lunatic*; he should know, for he has been so *struck* with his acquaintance, that he has adopted the man's name. Or ask any little girl in the nursery, and she will recite, with sweet simplicity, how

"The man in the moon
Came down too soon,
And asked the way to Norwich."

The darling may not understand why he sought that venerable

city, nor whether he ever arrived there, but she knows very well that

> "He went by the south,
> And burnt his mouth
> With eating hot pease porridge."

But it is useless to inquire of any stupid joker, for he will idly say that there is no such man there, because, forsooth, a certain single woman who was sent to the moon came back again, which she would never have done if a man had been there with whom she could have married and remained, Nor should any one be misled by those blind guides who darkly hint that it is all moonshine. There is not an Indian moonshee, nor a citizen of the Celestial Empire, some of whose ancestors came from the nocturnal orb, who does not know better than that. Perhaps the wisest course is to inquire within. Have not we all frequently affirmed that we knew no more about certain inscrutable matters than the man in the moon? Now we would never have committed ourselves to such a comparison had we not been sure that the said man was a veritable and creditable, though somewhat uninstructed person. But our feelings ought not to be wrought upon in this way. We "had rather be a dog, and bay the moon, than such a Roman" as is not at least distantly acquainted with that brilliant character in high life who careers so conspicuously amid the constellations which constitute the upper ten thousand of super-mundane society. And now some inquisitive individual may be impatient to interrupt our eloquence with the question, "What are you going to make of the man in the moon?" Well, we are not going to make anything of him. For, first, he is a man; therefore incapable of improvement. Secondly, he is in the moon, and that is out of our reach.* All that we can promise just now is, to furnish a few particulars of the man himself; some

* Besides, as old John Lilly says in the prologue to his *Endymion* (1591), "There liveth none under the sunne, that knows what to make of the man in the moone."

account of calls which he is reported to have made to his friends here below; and also some account of visits which his friends on earth have paid him in return.

We know something of his residence, whenever he is at home: what do we know of the man? We have been annoyed at finding his lofty name desecrated to base uses. If "imagination may trace the noble dust of Alexander, till he find it stopping a bung-hole," literature traces the man in the moon, and discovers him pressed into the meanest services. Our readers need not be disquieted with details; though our own equanimity has been sorely disturbed as we have seen scribblers dragging from the skies a "name at which the world grows pale, to point a moral, or adorn a tale." Political squibs, paltry chapbooks, puny satires, and penny imbecilities, too numerous for mention here, with an occasional publication of merit, have been printed and sold at the expense of the man in the moon. For the sake of the curious we place the titles and dates of some of these in an appendix and pass on. We have not learned very many particulars relating to the domestic habits or personal character of the man in the moon, consequently our smallest biographical contributions will be thankfully received. We must not be pressed for his photograph, at present. We certainly wish it could have been procured; but though photography has taken some splendid views of the face of the moon, it has not yet produced any perfect picture of the physiognomy of the man. It should always be borne in mind that, as Stilpo says in the old play of *Timon*, written about 1600, "The man in the moone is not in the moone superficially, although he bee in the moone (as the Greekes will have it) catapodially, specificatively, and quidditatively."[4] This beautiful language, let us explain for the behoof of any foreign reader, simply means that he is not always where we can get at him; and therefore his venerable visage is missing from our celestial portrait gallery. One fact we have found out, which we fear will ripple the pure water placidity of some of our best friends; but the truth must be told.

The Man in the Moon
Geo. Cruikshank. Hone's "*Facetiae*," 1821.

"If Caesar can hide the sun with a blanket, or put the moon in his pocket, we will pay him tribute for light" (*Cymbeline*).

"Our man in the moon drinks clarret,
With powder-beef, turnep, and carret.
If he doth so, why should not you
Drink until the sky looks blew?"[5]

Another old ballad runs:

"The man in the moon drinks claret,
But he is a dull Jack-a-Dandy;
Would he know a sheep's head from a carrot,
He should learn to drink cyder and brandy."

In a *Jest Book of the Seventeenth Century* we came across the following story: "A company of gentlemen coming into a tavern, whose signe was the Moone, called for a quart of sacke. The drawer told them they had none; whereat the gentlemen wondring were told by the drawer that the man in the moon always drunke claret."[6] Several astronomers assert the absence of water in the moon; if this be the case, what is the poor man to drink? Still, it is an unsatisfactory announcement to us all; for we are afraid that it is the claret which makes him look so red in the face sometimes when he is full, and gets a little fogged.

We have ourselves seen him actually what sailors call "half-seas over," when we have been in mid-Atlantic. We only hope that he imbibes nothing stronger, though it is said that moonlight is but another name for smuggled spirits. The lord of Cynthia must not be too hastily suspected, for, at most, the moon fills her horn but once a month. Still, the earth itself being so invariably sober, its satellite, like Caesar's wife, should be above suspicion. We therefore hope that our lunar hero may yet take a ribbon of sky-blue from the milky way, and become a staunch abstainer; if only for example's sake.

"The Man in the Moon Drinks Claret"
"*Bagford Ballads*," ii. 119.

Some old authors and artists have represented the man in the moon as an inveterate smoker, which habit surprises us, who supposed him to be

> "Above the smoke and stir of this dim spot
> Which men call Earth,"

as the magnificent Milton has it. His tobacco must be bird's-eye, as he takes a bird's-eye view of things; and his pipe is presumably a meer-sham, whence his "sable clouds turn forth their silver lining on the night." Smoking, without doubt, is a bad practice, especially when the clay is choked or the weed is worthless; but fuming against smokers we take to be infinitely worse.

We are better pleased to learn that the man in the moon is a poet. Possibly some uninspired groveller, who has never climbed

Parnassus, nor drunk of the Castalian spring, may murmur that this is very likely, for that all poetry is "moonstruck madness." Alas if such an antediluvian barbarian be permitted to "revisit thus the glimpses of the moon, making night hideous" as he mutters his horrid blasphemy! We, however, take a nobler view of the matter. To us the music of the spheres is exalting as it is exalted; and the music of earth is a "sphere-descended maid, friend of pleasure, wisdom's aid." We are therefore disposed to hear the following lines, which have been handed down for publication. Their title is autobiographical, and, for that reason, they are slightly egotistical.

Banks' Collection of Shop Bills

"A Shrewd Old Fellow's The Man in the Moon."

"From my palace of light I look down upon earth,
When the tiny stars are twinkling round me;
Though centuries old, I am now as bright
As when at my birth Old Adam found me.
Oh! the strange sights that I have seen,
Since earth first wore her garment of green!
King after king has been toppled down,
And red-handed anarchy's worn the crown!
From the world that's beneath me I crave not a boon,
For a shrewd old fellow's the Man in the Moon.
And I looked on 'mid the watery strife,
When the world was deluged and all was lost
Save one blessed vessel, preserver of life,
Which rode on through safety, though tempest tost.
I have seen crime clothed in ermine and gold,
And virtue shuddering in winter's cold.
I have seen the hypocrite blandly smile,
While straightforward honesty starved the while.
Oh! the strange sights that I have seen,
Since earth first wore her garment of green!
I have gazed on the coronet decking the brow
Of the villain who, breathing affection's vow,
Hath poisoned the ear of the credulous maiden,
Then left her to pine with heart grief laden.
Oh! oh! if this, then, be the world, say I,
I'll keep to my home in the clear blue sky;
Still to dwell in my planet I crave as a boon,
For the earth ne'er will do for the Man in the Moon."[7]

This effusion is not excessively flattering to our "great globe," and "all which it inherit"; and we surmise that the author was in a misanthropic mood when it was written. Yet it is serviceable sometimes to see ourselves as others see us. On the other hand, we have but little liking for those who "hope to merit heaven by making earth a hell," in any sense. We prefer to believe that the tide is rising though the waves recede, and that our dark world is waxing towards the full-orbed glory "to which the whole creation moves."

Here for the present we part company with the man in the moon as material for amusement, that we may track him through the mythic maze, where, in well-nigh every language, he has left some traces of his existence. As there is a side of the moon which we have never seen, and according to Laplace never shall see, there is also an aspect of the matter in hand that remains to be traversed, if we would circumambulate its entire extent. Our subject must now be viewed in the magic mirror of mythology. The antiquarian Ritson shall state the question to be brought before our honourable house of inquiry. He denominates the man in the moon "an imaginary being, the subject of perhaps one of the most ancient, as well as one of the most popular, superstitions of the world."[8] And as we must explore the vestiges of antiquity, Asiatic and European, African and American, and even Polynesian, we bespeak patient forbearance and attention. One little particular we may partly clear up at once, though it will meet us again in another connection. It will serve as a sidelight to our legendary scenes. In English, French, Italian, Latin, and Greek, the moon is feminine; but in all the Teutonic tongues the moon is masculine. Which of the twain is its true gender? We go back to the Sanskrit for an answer. Professor Max Müller rightly says, "It is no longer denied that for throwing light on some of the darkest problems that have to be solved by the student of language, nothing is so useful as a critical study of Sanskrit."[9] Here the word for the moon is *mâs*, which is masculine. Mark how even what Hamlet calls "words, words, words" lend their weight and value to the adjustment of this great argument. The very moon is masculine, and, like Wordsworth's child, is "father of the man."

If a bisexous moon seem an anomaly, perhaps the suggestion of Jamieson will account for the hermaphrodism: "The moon, it has been said, was viewed as of the masculine gender in respect of the earth, whose husband he was supposed to be; but as a female in relation to the sun, as being his spouse."[10] Here, also, we find a clue to the origin of this myth. If modern science, discovering the moon's inferiority to the sun, call the former feminine, ancient nescience, supposing the sun to be inferior to the moon, called the latter masculine. The sun, incomparable in splendour, invariable in aspect and motion, to the unaided eye immaculate in surface, too dazzling to permit prolonged observation, and shining in the daytime, when the mind was occupied with the duties of pastoral, agricultural, or commercial life, was to the ancient simply an object of wonder as a glory, and of worship as a god. The moon, on the contrary, whose mildness of lustre enticed attention, whose phases were an embodiment of change, whose strange spots seemed shadowy pictures of things and beings terrestrial, whose appearance amid the darkness of night was so welcome, and who came to men susceptible, from the influences of quiet and gloom, of superstitious imaginings, from the very beginning grew into a familiar spirit of kindred form with their own, and though regarded as the subordinate and wife of the sun, was reverenced as the superior and husband of the earth. With the transmission of this myth began its transmutation. From the moon being a man, it became a man's abode: with some it was the world whence human spirits came; with others it was the final home whither human spirits returned. Then it grew into a penal colony, to which egregious offenders were transported; or prison cage, in which, behind bars of light, miserable sinners were to be exposed to all eternity, as a warning to the excellent of the earth. One thing is certain, namely, that, during some phases, the moon's surface strikingly resembles a man's countenance. We usually represent the sun and the moon with the faces of men; and in the latter case the task is not difficult. Some would say that the moon is so drawn to reproduce some lunar deity: it would be more correct to say that the lunar deity was created through this human likeness. Sir Thomas Browne remarks, "The sun and moon are usually described with human faces: whether herein there be not a pagan imitation, and those visages at first implied Apollo and Diana, we may make some doubt."[11] Brand, in quoting Browne, adds, "Butler asks a shrewd question on this head, which I do not remember to have seen solved:—

"Tell me but what's the natural cause,
Why on a Sign no Painter draws
The *Full Moon* ever, but the *Half*?"

(Hudibras, B. II., c. iii.)[12]

Another factor in the formation of our moon-myth was the anthropomorphism which sees something manlike in everything, not only in the anthropoid apes, where we may find a resemblance more faithful than flattering, but also in the mountains and hills, rivers and seas of earth, and in the planets and constellations of heaven. Anthropomorphism was but a species of personification, which also metamorphosed the firmament into a menagerie of lions and bears, with a variety of birds, beasts, and fishes. Dr. Wagner writes: "The sun, moon, and stars, clouds and mists, storms and tempests, appeared to be higher powers, and took distinct forms in the imagination of man. As the phenomena of nature seemed to resemble animals either in outward form or in action, they were represented under the figure of animals."[13] Sir George W. Cox points out how phrases ascribing to things so named the actions or feelings of living beings, "would grow into stories which might afterwards be woven together, and so furnish the groundwork of what we call a legend or a romance. This will become plain, if we take the Greek sayings or myths about Endymion and Selênê. Here, besides these two names, we have the names Protogenia and Asterodia. But every Greek knew that Selênê was a name for the moon, which was also described as Asterodia because she has her path among the stars, and that Protogenia denoted the first or early born morning. Now Protogenia was the mother of Endymion, while Asterodia was his wife; and so far the names were transparent.

Had all the names remained so, no myth, in the strict sense of the word, could have sprung up; but as it so happened, the meaning of the name Endymion, as denoting the sun, when he is about to plunge or dive into the sea, had been forgotten, and thus Endymion became a beautiful youth with whom the moon fell in love, and whom she came to look upon as he lay in profound sleep in the cave of Latmos."[14] To this growth and transformation of myths we may return after awhile; meanwhile we will follow closely our man in the moon, who, among the Greeks, was the young Endymion, the beloved of Diana, who held the shepherd passionately in her embrace. This fable probably arose from Endymion's love of astronomy, a predilection common in ancient pastors. He was, no doubt, an ardent admirer of the moon; and soon it was reported that Selênê courted and caressed him in return. May such chaste enjoyment be ours also! We may remark, in passing, that classic tales are pure or impure, very much according to the taste of the reader. "To the jaundiced all things seem yellow," say the French; and Paul said, "To the pure all things are pure: but unto them that are defiled is nothing pure." According to Serapion, as quoted by Clemens Alexandrinus, the tradition was that the face which appears in the moon is the soul of a Sibyl. Plutarch, in his treatise, *Of the Face appearing in the roundle of the Moone*, cites the poet Agesinax as saying of that orb,

"All roundabout environed
With fire she is illumined:
And in the middes there doth appeere,
Like to some boy, a visage cleere;
Whose eies to us doe seem in view,
Of colour grayish more than blew:
The browes and forehead tender seeme,
The cheeks all reddish one would deeme."[15]

The story of the man in the moon as told in our British nurseries is supposed to be founded on Biblical fact. But though the Jews have a Talmudic tradition that Jacob is in the moon, and though they believe that his face is plainly visible, the Hebrew Scriptures make no mention of the myth. Yet to our fireside auditors it is related that a man was found by Moses gathering sticks on the Sabbath, and that for this crime he was transferred to the moon, there to remain till the end of all things. The passage cited in support of this tale is *Numbers* xv. 32-36. Upon referring to the sacred text, we certainly find a man gathering sticks upon the Sabbath day, and the congregation gathering stones for his merciless punishment, but we look in vain for any mention of the moon. *Non est inventus.* Of many an ancient story-teller we may say, as Sheridan said of Dundas, "the right honourable gentleman is indebted to his memory for his jests and to his imagination for his facts."

Mr. Proctor reminds us that "according to German nurses, the day was not the Sabbath, but Sunday. Their tale runs as follows: Ages ago there went one Sunday an old man into the woods to hew sticks. He cut a faggot and slung it on a stout staff, cast it over his shoulder, and began to trudge home with his burthen. On his way he met a handsome man in Sunday suit, walking towards the church. The man stopped, and asked the faggot-bearer, 'Do you know that this is Sunday on earth, when all must rest from their labours?' 'Sunday on earth, or Monday in heaven, it's all one to me!' laughed the woodcutter. 'Then bear your bundle for ever!' answered the stranger. 'And as you value not Sunday on earth, yours shall be a perpetual moon-day in heaven; you shall stand for eternity in the moon, a warning to all Sabbath-breakers.' Thereupon the stranger vanished, and the man was caught up with his staff and faggot into the moon, where he stands yet."[16]

In Tobler's account the man was given the choice of burning in the sun, or of freezing in the moon; and preferring a lunar frost to a solar furnace, he is to be seen at full moon seated with his bundle of sticks on his back. If "the cold in clime arc cold in blood," we may be thankful that we do not hibernate eternally in the moon and in the nights of winter, when the cold north winds blow," we may look up through the casement and "pity the sorrows of this poor old man."

Mr. Baring-Gould finds that "in Schaumberg-lippe, the story goes, that a man and a woman stand in the moon: the man because he strewed brambles and thorns on the church path, so as to hinder people from attending mass on Sunday morning; the woman because she made butter on that day. The man carries his bundle of thorns, the woman her butter tub. A similar tale is told in Swabia and in Marken. Fischart says that there 'is to be seen in the moon a mannikin who stole wood'; and Praetorius, in his description of the world, that 'superstitious people assert that the black flecks in the moon are a man who gathered wood on a Sabbath, and is therefore turned into stone.'"[17]

The North Frisians, among the most ancient and pure of all the German tribes, tell the tale differently. "At the time when wishing was of avail, a man, one Christmas Eve, stole cabbages from his neighbour's garden. When just in the act of walking off with his load, he was perceived by the people, who conjured (wished) him up in the moon. There he stands in the full moon, to be seen by everybody, bearing his load of cabbages to all eternity. Every Christmas Eve he is said to turn round once. Others say that he stole willow-boughs, which he must bear for ever. In Sylt the story goes that he was a sheep-stealer, that enticed sheep to him with a bundle of cabbages, until, as an everlasting warning to others, he was placed in the moon, where he constantly holds in his hand a bundle of cabbages. The people of Rantum say that he is a giant, who at the time of the flow stands in a stooping posture, because he is then taking up water, which he pours out on the earth, and thereby causes the flow; but at the time of the ebb he stands erect and rests from his labour, when the water can subside again."[18]

Crossing the sea into Scandinavia, we obtain some valuable information. First, we find that in the old Norse, or language of the ancient Scandinavians, the sun is always feminine, and the moon masculine. In the *Völu-Spá*, a grand, prophetic poem, it is written—

> "But the sun had not yet learned to trace
> The path that conducts to her dwelling-place
> To the moon arrived was not the hour
> When he should exert his mystic power
> Nor to the stars was the knowledge given,
> To marshal their ranks o'er the fields of heaven."[19]

We also learn that "the moon and the sun are brother and sister; they are the children of Mundilföri, who, on account of their beauty, called his son Mâni, and his daughter Sôl." Here again we observe that the moon is masculine. "Mâni directs the course of the moon, and regulates Nyi (the new moon) and Nithi (the waning moon). He once took up two children from the earth, Bil and Hiuki, as they were going from the well of Byrgir, bearing on their shoulders the bucket Soeg, and the pole Simul."[20] These two children, with their pole and bucket, were placed in the moon, "where they could be seen from earth"; which phrase must refer to the lunar spots. Thorpe, speaking of the allusion in the *Edda* to these spots, says that they "require but little illustration. Here they are children carrying water in a bucket, a superstition still preserved in the popular belief of the Swedes."[21] We are all reminded at once of the nursery rhyme—

"Jack and Jill went up the hill,
 To fetch a pail of water;
Jack fell down and broke his crown,
 And Jill came tumbling after."

Little have we thought, when rehearsing this jingle in our juvenile hours, that we should some day discover its roots in one of the oldest mythologies of the world. But such is the case. Mr. Baring-Gould has evolved the argument in a manner which, if not absolutely conclusive in each point, is extremely cogent and clear. "This verse, which to us seems at first sight nonsense, I have no hesitation in saying has a high antiquity, and refers to the Eddaic Hjuki and Bil. The names indicate as much. Hjuki, in Norse, would be pronounced Juki, which would readily become Jack; and Bil, for the sake of euphony and in order to give a female name to one of the children, would become Jill. The fall of Jack, and the subsequent fall of Jill, simply represent the vanishing of one moon spot after another, as the moon wanes. But the old Norse myth had a deeper signification than merely an explanation of the moon spots. Hjuki is derived from the verb jakka, to heap or pile together, to assemble and increase; and Bil, from bila, to break up or dissolve. Hjuki and Bil, therefore, signify nothing more than the waxing and waning of the moon, and the water they are represented as bearing signifies the fact that the rainfall depends on the phases of the moon. Waxing and waning were individualized, and the meteorological fact of the connection of the rain with the moon was represented by the children as water-bearers. But though Jack and Jill became by degrees dissevered in the popular mind from the moon, the original myth went through a fresh phase, and exists still under a new form. The Norse superstition attributed *theft* to the moon, and the vulgar soon began to believe that the figure they saw in the moon was the thief. The lunar specks certainly may be made to resemble one figure, but only a lively imagination can discern two. The girl soon dropped out of popular mythology, the boy oldened into a venerable man, he retained his pole, and the bucket was transformed into the thing he had stolen—sticks or vegetables. The theft was in some places exchanged for Sabbath-breaking, especially among those in Protestant countries who were acquainted with the Bible story of the stick-gatherer."[22]

The German Grimm, who was by no means a grim German, but a very genial story-teller, also maintains this transformation of the original myth. "Plainly enough the water-pole of the heathen story has been transformed into the axe's shaft, and the carried pail into the thornbush; the general idea of theft was retained, but special stress laid on the keeping of the Christian holiday, the man suffers punishment not so much for cutting firewood, as because he did it on a Sunday."[23] Manifestly "Jack and Jill went up the hill" is more than a Runic rhyme, and like many more of our popular strains might supply us with a most interesting and instructive entertainment; but we must hasten on with the moon-man.

We come next to Britain. Alexander Neckam, a learned English abbot, poet, and scholar, born in St. Albans, in 1157, in commenting on the dispersed shadow in the moon, thus alluded to the vulgar belief: "Nonne novisti quid vulgus vocet rusticum in luna portantem spinas? Unde quidam vulgariter loquens ait,

Rusticus in Luna
Quem sarcina deprimit una
Monstrat per spinas
Nulli prodesse rapinas."[24]

This may be rendered, "Do you not know what the people call the rustic in the moon who carries the thorns? Whence one vulgarly speaking says,

The Rustic in the moon,
Whose burden weighs him down,
This changeless truth reveals,
He profits not who steals."

Thomas Wright considers Neckam's Latin version of this popular distich "very curious, as being the earliest allusion we have to the popular legend of the man in the moon." We are specially struck with the reference to theft; while no less noteworthy is the absence of that sabbatarianism, which is the "moral" of the nursery tale.

In the British Museum there is a manuscript of English poetry of the thirteenth century, containing an old song composed probably about the middle of that century. It was first printed by Ritson in his *Ancient Songs*, the earliest edition of which was published in London, in 1790. The first lines are as follows:

> "Mon in the mone stond ant strit,
> On is bot-forke is burthen he bereth
> Hit is muche wonder that he na down slyt,
> For doute leste he valle he shoddreth and shereth."[25]

In the *Archaeological Journal* we are presented with a relic from the fourteenth century. "Mr. Hudson Taylor submitted to the Committee a drawing of an impression of a very remarkable personal seal, here represented of the full size. It is appended to a deed (preserved in the Public Record Office) dated in the ninth year of Edward the Third, whereby Walter de Grendene, clerk, sold to Margaret, his mother, one messuage, a barn and four acres of ground in the parish of Kingston-on-Thames. The device appears to be founded on the ancient popular legend that a husbandman who had stolen a bundle of thorns from a hedge was, in punishment of his theft, carried up to the moon. The legend reading *Te Waltere docebo cur spinas phebo gero*, 'I will teach you, Walter, why I carry thorns in the moon,' seems to be an enigmatical mode of expressing the maxim that honesty is the best policy."[26]

About fifty years later, in the same century, Geoffrey Chaucer, in his *Troylus and Creseide* adverts to the subject in these lines:

"(Quod Pandarus) Thou hast a full great care
Lest the chorl may fall out of the moone."

(Book i. Stanza 147.)

And in another place he says of Lady Cynthia, or the moon:

"Her gite was gray, and full of spottis blake,
And on her brest a chorl painted ful even,
Bering a bush of thornis on his backe,
Whiche for his theft might clime so ner the heaven."

Whether Chaucer wrote the *Testament and Complaint of Creseide*, in which these latter lines occur, is doubted, though it is frequently ascribed to him.[27]

Dr. Reginald Peacock, Bishop of Chichester, in his *Repressor*, written about 1449, combats "this opinioun, that a man which stale sumtyme a birthan of thornis was sett in to the moone, there for to abide for euere."

Thomas Dekker, a British dramatist, wrote in 1630: "A starre? Nay, thou art more than the moone, for thou hast neither changing quarters, nor a man standing in thy circle with a bush of thornes."[28]

And last, but not least, amid the tuneful train, William Shakespeare, without whom no review of English literature or of poetic lore could be complete, twice mentions the man in the moon. First, in the *Midsummer Night's Dream*, Act iii. Scene 1, Quince the carpenter gives directions for the performance of Pyramus and Thisby, who "meet by moonlight," and says, "One must come in with a bush of thorns and a lanthorn, and say he comes to disfigure, or to present, the person of Moonshine." Then in Act v. the player of that part says, "All that I have to say is, to tell you that the lanthorn is the moon; I, the man in the moon; this thorn-bush, my thorn-bush; and this dog, my dog." And, secondly, in the *Tempest*, Act ii., Scene 2, Caliban and Stephano in dialogue:

"*Cal.* Hast thou not dropp'd from heaven?
Ste. Out o' the moon, I do assure thee. I was the man i' the moon, when time was.
Cal. I have seen thee in her, and I do adore thee: my mistress show'd me thee, thy dog, and bush."

Robert Chambers refers the following singular lines to the man in the moon: adding, "The allusion to Jerusalem pipes is curious; Jerusalem is often applied, in Scottish popular fiction, to things of a nature above this world":

"I sat upon my houtie croutie (hams),
I lookit owre my rumple routie (haunch),
And saw John Heezlum Peezlum
Playing on Jerusalem pipes."[29]

Here is an old-fashioned couplet belonging probably to our northern borders:

"The man in the moon
Sups his sowins with a cutty spoon."

Halliwell explains *sowins* to be a Northumberland dish of coarse oatmeal and milk, and a *cutty* spoon to be a very *small* spoon.[30]

Wales is not without a memorial of this myth, for Mr. Baring-

Gould tells us that "there is an ancient pictorial representation of our friend the Sabbath-breaker in Gyffyn Church, near Conway. The roof of the chancel is divided into compartments, in four of which are the evangelistic symbols, rudely, yet effectively painted. Besides these symbols is delineated in each compartment an orb of heaven. The sun, the moon, and two stars, are placed at the feet of the Angel, the Bull, the Lion, and the Eagle. The representation of the moon is as follows: in the disk is the conventional man with his bundle of sticks, but without the dog."[31] Mr. Gould says, "our friend the Sabbath-breaker" perhaps the artist would have said "the thief," for stealing appears to be more antique.

Representation in Gyffyn Church, Near Conway.

A French superstition, lingering to the present day, regards the man in the moon as Judas Iscariot, transported to the moon for his treason. This plainly is a Christian invention. Some say the figure is Isaac bearing a burthen of wood for the sacrifice of himself on Mount Moriah. Others that it is Cain carrying a bundle of thorns on his shoulder, and offering to the Lord the cheapest gift from the field.[32] This was Dante's view, as the succeeding passages will show:

"For now doth Cain with fork of thorns confine
On either hemisphere, touching the wave
Beneath the towers of Seville. Yesternight
The moon was round."

(*Hell.* Canto xx., line 123.)

"But tell, I pray thee, whence the gloomy spots
Upon this body, which below on earth
Give rise to talk of Cain in fabling quaint?"

(*Paradise*, ii. 50.)[33]

When we leave Europe, and look for the man in the moon under other skies, we find him, but with an altogether new aspect. He is the same, and yet another; another, yet the same. In China he plays a pleasing part in connubial affairs. "The Chinese 'Old Man in the Moon' is known as *Yue-lao*, and is reputed to hold in his hands the power of predestining the marriages of mortals—so that marriages, if not, according to the native idea, exactly made in heaven, are made somewhere beyond the bounds of earth. He is supposed to tie together the future husband and wife with an invisible silken cord, which never parts so long as life exists."[34] This must be the man of the Honey-moon, and we shall not meet his superior in any part of the world. Among the Khasias of the Himalaya Mountains "the changes of the moon are accounted for by the theory that this orb, who is a man, monthly falls in love with his wife's mother, who throws ashes in his face. The sun is female."[35] The Slavonic legend, following the Himalayan, says that "the moon, King of night and husband of the sun, faithlessly loves the morning Star, wherefore he was cloven through in punishment, as we see him in the sky."[36]

"One man in his time plays many parts," and the man in the moon is no exception to the rule. In Africa his *rôle* is a trying one; for "in Bushman astrological mythology the moon is looked upon as a man who incurs the wrath of the sun, and is consequently pierced by the knife (*i.e.* rays) of the latter. This process is repeated until almost the whole of the moon is cut away, and only a little piece left; which the moon piteously implores the sun to spare for his (the moon's) children. (The moon is in Bushman mythology a male being.) From this little piece, the moon gradually grows again until it becomes a full moon, when the sun's stabbing and cutting processes recommence."[37]

We cross the Atlantic, and among the Greenlanders discover a myth, which is *sui generis*. "The sun and moon are nothing else than two mortals, brother and sister. They were playing with others at children's games in the dark, when *Malina*, being teased in a shameful manner by her brother *Anninga*, smeared her hands with the soot of the lamp, and rubbed them over the face and hands of her persecutor, that she might recognise him by daylight. Hence arise the spots in the moon. Malina wished to save herself by flight, but her brother followed at her heels. At length she flew upwards, and became the sun. Anninga followed her, and became the moon; but being unable to mount so high, he runs continually round the sun, in hopes of some time surprising her. When he is tired and hungry in his last quarter, he leaves his house on a sledge harnessed to four huge dogs, to hunt seals, and continues abroad for several days. He now fattens so prodigiously on the spoils of the chase, that he soon grows into the full moon. He rejoices on the death of women, and the sun has her revenge on the death of men; all males therefore keep within doors during an eclipse of the sun, and females during that of the moon."[38] This Esquimaux story, which has some interesting features, is told differently by Dr. Hayes, the Arctic explorer, who puts a lighted taper into the sun's hands, with which she discovered her brother, and which now causes her bright light, "while the moon, having lost his taper, is cold, and could not be seen but for his sister's light."[39] This belief prevails as far south as Panama, for the inhabitants of the Isthmus of Darien have a tradition that the man in the moon was guilty of gross misconduct towards his elder sister, the sun.[40]

The Creek Indians say that the moon is inhabited by a man and a dog. The native tribes of British Columbia, too, have their myth.

Mr. William Duncan writes to the Church Missionary Society: "One very dark night I was told that there was a moon to be seen on the beach. On going to see, there was an illuminated disk, with the figure of a man upon it. The water was then very low, and one of the conjuring parties had lit up this disk at the water's edge. They had made it to wax with great exactness, and presently it was at full. It was an imposing sight. Nothing could be seen around it; but the Indians suppose that the medicine party are then holding converse with the man in the moon."[41] Mr. Duncan was at another time led to the ancestral village of a tribe of Indians, whose chief said to him: "This is the place where our fore fathers lived, and they told us something we want to tell you. The story is as follows: 'One night a child of the chief class awoke and cried for water. Its cries were very affecting—"Mother, give me to drink!" but the mother heeded not. The moon was affected, and came down, entered the house, and approached the child, saying, "Here is water from heaven: drink." The child anxiously laid hold of the pot and drank the draught, and was enticed to go away with the moon, its benefactor. They took an underground passage till they got quite clear of the village, and then ascended to heaven.' And," said the chief, "our forefathers tell us that the figure we now see in the moon is that very child; and also the little round basket which it had in its hand when it went to sleep appears there."[42]

The aborigines of New Zealand have a suggestive version of this superstition. It is quoted from D'Urville by De Rougemont in his *Le Peuple Primitif* (tom. ii. p. 245), and is as follows:—"Before the moon gave light, a New Zealander named Rona went out in the night to fetch some water from the well. But he stumbled and unfortunately sprained his ankle, and was unable to return home. All at once, as he cried out for very anguish, he beheld with fear and horror that the moon, suddenly becoming visible, descended towards him. He seized hold of a tree, and clung to it for safety; but it gave way, and fell with Rona upon the moon; and he remains there to this day."[43] Another account of Rona varies in that he escapes falling into the well by seizing a tree, and both he and the tree were caught up to the moon. The variation indicates that the legend has a living root.

Here we terminate our somewhat wearisome wanderings about the world and through the mazes of mythology in quest of the man in the moon. As we do so, we are constrained to emphasize the striking similarity between the Scandinavian myth of Jack and Jill, that exquisite tradition of the British Columbian chief, and the New Zealand story of Rona. When three traditions, among peoples so far apart geographically, so essentially agree in one, the lessons to be learned from comparative mythology ought not to be lost upon the philosophical student of human history. To the believer in the unity of our race such a comparison of legends is of the greatest importance. As Mr. Tylor tells us, "The number of myths recorded as found in different countries, where it is hardly conceivable that they should have grown independently, goes on steadily increasing from year to year, each one furnishing a new clue by which common descent or intercourse is to be traced."[44] The same writer says on another page of his valuable work, "The mythmaking faculty belongs to mankind in general, and manifests itself in the most distant regions, where its unity of principle develops itself in endless variety of form."[45] Take, for example, China and England, representing two distinct races, two languages, two forms of religion, and two degrees of civilization yet, as W. F. Mayers remarks, "No one can compare the Chinese legend with the popular European belief in the 'Man in the Moon,' without feeling convinced of the certainty that the Chinese superstition and the English nursery tale are both derived from kindred parentage, and are linked in this relationship by numerous subsidiary ties. In all the range of Chinese mythology there is, perhaps, no stronger instance of identity with the traditions that have taken root in Europe than in the case of the legends relating to the moon."[46] This being the case, our present endeavour to establish the consanguinity of the nations, on the ground of agreement in myths and modes of faith and worship, cannot be labour thrown away. The recognition of friends in heaven is an interesting speculation; but far more good must result, as concerns this life at least, from directing our attention to the recognition of friends on earth. If we duly estimate the worth of any comparative science, whether of anatomy or philology, mythology or religion, this is the grand generalization to be attained, essential unity consistent and concurrent with endless multiformity; many structures, but one life; many creeds, but one faith; many beings and becomings, but all emanating from one Paternity, cohering through one Presence, and converging to one Perfection, in Him who is the Author and Former and Finisher of all things which exist. Let no man therefore ridicule a myth as puerile if it be an aid to belief in that commonweal of humanity for which the Founder of the purest religion was a witness and a martyr. We have sought out the man in the moon mainly because it was one out of many scattered stories which, as Max Müller nobly says, "though they may be pronounced childish and tedious by some critics, seem to me to glitter with the brightest dew of nature's own poetry, and to contain those very touches that make us feel akin, not only with Homer or Shakespeare, but even with Lapps, and Finns, and Kaffirs."[47] Vico discovered the value of myths, as an addition to our knowledge of the mental and moral life of the men of the myth-producing period. Professor Flint tells us that mythology, as viewed by the contemporaries of Vico, "appeared to be merely a rubbish-heap, composed of waste, worthless, and foul products of mind; but he perceived that it contained the materials for a science which would reflect the mind and history of humanity, and even asserted some general principles as to how these materials were to be interpreted and utilised, which have since been established, or at least endorsed, by Heyne, Creuzer, C O Müller, and others."[48] Let us cease to call that common which God has cleansed, and with thankfulness recognise the solidarity of the human race, to which testimony is borne by even a lunar myth.

We now return to the point whence we deflected, and rejoin the chief actor in the selenographic comedy. It is a relief to get away from the legendary man in the moon, and to have the real man once more in sight. We are like the little boy, whom the obliging visitor, anxious to show that he was passionately fond of children, and never annoyed by them in the least, treated to a ride upon his knee. "Trot, trot, trot; how do you enjoy that, my little man? Isn't that nice?" "Yes, sir," replied the child, "but not so nice as on the real donkey, the one with the four legs." It is true, the mythical character has redeeming traits; but then he breaks the Sabbath, obstructs people going to mass, steals cabbages, and is undergoing sentence of transportation for life. While the real man, who lives in a well-lighted crescent, thoroughly ventilated; whose noble profile is sometimes seen distinctly when he passes by on the shady side of the way; whose beaming countenance is at other times turned full upon us, reflecting nothing but sunshine as he winks at his many admirers: he is a being of quite another order. We do not forget that he has been represented with a claret jug in one hand, and a claret cup in the other; that he frequently takes half and half; that he is a smoker; that he sometimes gets up when other people are going to bed; that he often stops out all the night; and is too familiar with the low song—

> "We won't go home till morning."

But these are mere eccentricities of greatness, and with all such irregularities he is "a very delectable, highly respectable" young fellow; in short,

> "A most intense young man,
> A soul-full-eyed young man,
> An ultra-poetical, super-aesthetical,
> Out-of-the-way young man."

Why, he has been known to take the shine out of old Sol himself; though from his partiality to us it always makes him look black in the face when we, Alexander-like, stand between him and that luminary. We, too, are the only people by whom he ever allows himself to be eclipsed. Illustrious man in the moon I he has lifted our thoughts from earth to heaven, and we are reluctant to leave him. But the best of friends must part; especially as other lunar inhabitants await attention.

"Other inhabitants!" some one may exclaim. Surely! we reply; and though it will necessitate a digression, we touch upon the question *en passant*. Cicero informs us that "Xenophanes says that the moon is inhabited, and a country having several towns and mountains in it."[49] This single dictum will be sufficient for those who bow to the influence of authority in matters of opinion. Settlement of questions by "texts" is a saving of endless pains. For that there are such lunar inhabitants must need little proof. Every astronomer is aware that the moon is full of craters; and every linguist is aware that "cratur" is the Irish word for creature. Or, to state the argument syllogistically, as our old friend Aristotle would have done: "Craturs" are inhabitants; the moon is full of craters; therefore the moon is full of inhabitants. We appeal to any unbiased mind whether such argumentation is not as sound as much of our modern reasoning, conducted with every pretence to logic and lucidity. Besides, who has not heard of that astounding publication, issued fifty years since, and entitled *Great Astronomical Discoveries lately made by Sir John Herschel, LL.D., F.R.S., etc., at the Cape of Good Hope*? One writer dares to designate it a singular satire; stigmatizes it as the once celebrated *Moon Hoax*, and attributes it to one Richard Alton Locke, of the United States. What an insinuation! that a man born under the star-spangled banner could trifle with astronomy. But if a few incredulous persons doubted, a larger number of the credulous believed. When the first number appeared in the New York Sun, in September, 1835, the excitement aroused was intense. The paper sold daily by thousands; and when the articles came out as a pamphlet, twenty thousand went off at once. Not only in Young America, but also in Old England, France, and throughout Europe, the wildest enthusiasm prevailed. Could anybody reasonably doubt that Sir John had seen wonders, when it was known that his telescope contained a prodigious lens, weighing nearly seven tons, and possessing a magnifying power estimated at 42,000 times? A reverend astronomer tells us that Sir Frederick Beaufort, having occasion to write to Sir John Herschel at the Cape, asked if he had heard of the report current in England that he (Sir John) had discovered sheep, oxen, and flying *men* in the moon. Sir John had heard the report; and had further heard that an American divine had "improved" the revelations. The said divine had told his congregation that, on account of the wonderful discoveries of the present age, lie lived in expectation of one day calling upon them for a subscription to buy Bibles for the benighted inhabitants of the moon.[50] What more needs to be said? Give our astronomical mechanicians a little time, and they will produce an instrument for full verification of these statements regarding the lunar inhabitants; and we may realize more than we have imagined or dreamed. We may obtain observations as satisfactory as those of a son of the Emerald Isle, who was one day boasting to a friend of his excellent telescope. "Do you see yonder church?" said he. "Although it is scarcely discernible with the naked eye, when I look at it through my telescope, it brings it so close that I can hear the organ playing." Two hundred years ago, a wise man witnessed a wonderful phenomenon in the moon: he actually beheld a live elephant there. But the unbelieving have ever since made all manner of fun at the good knight's expense. Take the following burlesque of this celebrated discovery as an instance. "Sir Paul Neal, a conceited virtuoso of the seventeenth century, gave out that he had discovered 'an elephant in the moon.' It turned out that a mouse had crept into his telescope, which had been mistaken for an elephant in the moon."[51] Well, we concede that an elephant and a mouse are very much alike; but surely Sir Paul was too sagacious to be deceived by resemblances. If we had more faith, which is indispensable in such matters, the revelations of science, however extraordinary or extravagant, would be received without a murmur of distrust. We should not then meet with such sarcasm as we found in the seventeenth century *Jest Book* before quoted: "One asked why men should thinke there was a world in the moone? It was answered, because they were lunatique."

According to promise, we must make mention of at least one visit paid by our hero to this lower world. We do this in the classic language of a student of that grand old University which stands in the city of Oxford. May the horns of Oxford be exalted, and the shadow of the University never grow less, while the moon endureth!

"The man in the moon! why came he down
From his peaceful realm on high;
Where sorrowful moan is all unknown,
And nothing is born to die?
The man in the moon was tired, it seems,
Of living so long in the land of dreams;
'Twas a beautiful sphere, but nevertheless
Its lunar life was passionless;
Unchequered by sorrow, undimmed by crime,
Untouched by the wizard wand of time;
'Twas all too grand, there was no scope
For dread, and of course no room for hope
To him the future had no fear,
To make the present doubly dear;
The day no cast of coming night,
To make the borrowed ray more bright;
And life itself no thought of death,
To sanctify the boon of breath:—
In short, as we world-people say,
The man in the moon was *ennuyé*."[52]

Poor man in the moon! what a way he must have been in! We hope that he found improving fellowship, say among the Fellows of some Royal Astronomical Society; and that when e returned to his skylight, or lighthouse on the coast of immensity's wide sea, he returned a wiser and much happier man. It is for us, too, to remember with Spenser, "The noblest mind the best contentment has."

And now we record a few visits which men of this sublunary sphere are said to have paid to the moon. The chronicles are unfortunately very incomplete. Aiming at historical fulness and fidelity, we turned to our national bibliotheca at the British Museum, where we fished out of the vasty deep of treasures a MS. without date or name. We wish the Irish orator's advice were oftener followed by literary authors. Said he, "Never write an anonymous letter without signing your name to it." This MS. is entitled "*Selenographia*, or News from the world in the moon to the lunatics of this world. By Lucas Lunanimus of Lunenberge."[53] We are here told how the author, "making himself a kite of ye hight(?) of a large sheet, and tying himself to the tayle of it, by the help of some trusty friends, to whom he promised mountains of land in this his new-found world; being furnished also with a tube, horoscope, and other instruments of discovery, he set saile the first of Aprill, a day alwaies esteemed prosperous for such adventures." Fearing, however, lest the date of departure should make some suspicious that the author was desirous of making his readers April fools, we leave this aërial tourist to pursue his explorations without our company, and listen to a learned bishop, who ought to be a canonical authority, for the man in the moon himself is an overseer of men. Dr. Francis Godwin, first of Llandaff, afterwards of Hereford, wrote about the year 1600 *The Man in the Moone*, or a discourse of a voyage thither. This was published in 1638, under the pseudonym of Domingo Gonsales. The enterprising aeronaut went up from the island of El Pico, carried by wild swans. *Swans*, be it observed. It was not a wild-goose chase. The author is careful to tell us what we believe so soon as it is declared. "The further we went, the lesser the globe of the earth appeared to us; whereas still on the contrary side the moone showed herselfe more and more monstrously huge." After eleven days' passage, the exact time that Arago allowed for a cannon ball to reach the moon, "another earth" was approached. "I perceived that it was covered for the most part with a huge and mighty sea, those parts only being drie land, which show unto us here somewhat darker than the rest of her body; that I mean which the country people call *el hombre della Luna*, the man of the moone." This last clause demands a protest. The bishop knocks the country-people's man out of the moon, to make room for his own man, which episcopal creation is twenty-eight feet high, and weighs twenty-five or thirty of any of us. Besides ordinary men, of extraordinary measurement, the bishop finds in the moon princes and queens. The females, or lunar ladies, as a matter of course, are of absolute beauty. Their language has "no affinity with any other I ever heard." This is a poor look-out for the American divine who expects to send English Bibles to the moon. "Food groweth everywhere without labour": this is a cheering prospect for our working classes who may some day go there. "They need no lawyers": oh what a country! "And as little need is there of physicians." Why, the moon must be Paradise regained. But, alas! "they die, or rather (I should say) cease to live." Well, my lord bishop, is not that how we die on earth? Perhaps we need to be learned bishops to appreciate the difference. If so, we might accept episcopal distinction.

Lucian, the Greek satirist, in his *Voyage to the Globe of the Moon*, sailed through the sky for the space of seven days and nights and on the eighth "arrived in a great round and shining island which hung in the air and yet was inhabited. These inhabitants were Hippogypians, and their king was Endymion."[54] Some of the ancients thought the lunarians were fifteen times larger than we are, and our oaks but bushes compared with their trees. So natural is it to magnify prophets not of our own country.

William Hone tells us that a Mr. Wilson, formerly curate of Halton Gill, near Skipton-in-Craven, Yorkshire, in the last century wrote a tract entitled *The Man in the Moon*, which was seriously meant to convey the knowledge of common astronomy in the following strange vehicle: A cobbler, Israel Jobson by name, is supposed to ascend first to the top of Penniguit; and thence, as a second stage equally practicable, to the moon; after which he makes the grand tour of the whole solar system. From this excursion, however, the traveller brings back little information which might not have been had upon earth, excepting that the inhabitants of one of the planets, I forget which, were made of "pot metal."[55] This curious tract, full of other extravagances, is rarely if ever met with, it having been zealously bought up by its writer's family.

We must not be detained with any detailed account of M. Jules Verne's captivating books, entitled *From the Earth to the Moon*, and *Around the Moon*. They are accessible to all, at a trifling cost. Besides, they reveal nothing new relating to the Hamlet of our present play. Nor need we more than mention "the surprising adventures of the renowned Baron Munchausen." His lunarians being over thirty-six feet high, and "a common flea being much larger than one of our sheep,"[56] Munchausen's moon must be declined, with thanks.

"Certain travellers, like the author of the *Voyage au monde de Descartes*, have found, on visiting these different lunar countries, that the great men whose names they had arbitrarily received took possession of them in the course of the sixteenth century, and there fixed their residence. These immortal souls, it seems, continued their works and systems inaugurated on earth. Thus it is, that on Mount Aristotle a real Greek city has risen, peopled with peripatetic philosophers, and guarded by sentinels armed with propositions, antitheses, and sophisms, the master himself living in the centre of the town in a magnificent palace. Thus also in Plato's circle live souls continually occupied in the study of the prototype of ideas. Two years ago a fresh division of lunar property was made, some astronomers being generously enriched."[57]

That the moon is an abode of the departed spirits of men, an upper hades, has been believed for ages. In the Egyptian *Book of Respirations*, which M. p. J. de Horrack has translated from the MS. in the Louvre in Paris, Isis breathes the wish for her brother Osiris "that his soul may rise to heaven in the disk of the moon."[58] Plutarch says, "Of these soules the moon is the element, because soules doe resolve into her, like as the bodies of the dead into the earth."[59] To this ancient theory Mr. Tylor refers when he writes, "And when in South America the Saliva Indians have pointed out the moon, their paradise where no mosquitoes are, and the Guaycurus have shown it as the home of chiefs and medicine-men deceased, and the Polynesians of Tokelau in like manner have claimed it as the abode of departed kings and chiefs, then these pleasant fancies may be compared with that ancient theory mentioned by Plutarch, that hell is in the air and elysium in the moon, and again with the mediaeval conception of the moon as the seat of hell, a thought elaborated in profoundest bathos by Mr. M. F. Tupper:

'I know thee well, O Moon, thou cavern'd realm,
Sad satellite, thou giant ash of death,
Blot on God's firmament, pale home of crime,
Scarr'd prison house of sin, where damnèd souls
Feed upon punishment. Oh, thought sublime,
That amid night's black deeds, when evil prowls
Through the broad world, thou, watching sinners well,
Glarest o'er all, the wakeful eye of—Hell!'

Skin for skin, the brown savage is not ill-matched in such speculative lore with the white philosopher."[60]

The last journey to the moon on our list we introduce for the sake of its sacred lesson. Pure religion is an Attic salt, which wise men use in all of their entertainments: a condiment which seasons what is otherwise insipid, and assists healthy digestion in the compound organism of man's mental and moral constitution. About seventy years since, a little tract was published, in which the writer imagined himself on *luna firma*. After giving the inhabitants of the moon an account of our terrestrial race, of its fall and redemption, and of the unhappiness of those who neglect the great salvation, he says, "The secret is this, that nothing but an infinite God, revealing Himself by His Spirit to their minds, and enabling them to believe and trust in Him, can give perfect and lasting satisfaction." He then adds, "My last observation received the most marked approbation of the lunar inhabitants: they truly pitied the ignorant triflers of our sinful world, who prefer drunkenness, debauchery, sinful amusements, exorbitant riches, flattery, and other things that are highly esteemed amongst men, to the pleasures of godliness, to the life of God in the soul of man, to the animating hope of future bliss."[61]

Here the man in the moon and we must part. Hitherto some may have supposed their thoughts occupied with a mere creature of imagination, or gratuitous creation of an old-world mythology. Perhaps the man in the moon is nothing more: perhaps he is very much more. Possibly we have information of every being in the universe; and possibly there are beings in every existing world of which we know nothing whatever. The latter possibility we deem much the more probable. Remembering our littleness as contrasted with the magnitude of the whole creation, we prefer to believe that there are rational creatures in other worlds besides this small-sized sphere in, it may be, a small-sized system. Therefore, till we acquire more conclusive evidence than has yet been adduced, we will not regard even the moon as an empty abode, but as the home of beings whom, in the absence of accurate definition, we denominate men. Whether the man in the moon have a body like our own, whether his breathing apparatus, his digestive functions, and his cerebral organs, be identical with ours, are matters of secondary moment. The Fabricator of terrestrial organizations has limited himself to no one type or form, why then should man be the model of beings in distant worlds? Be the man in the moon a biped or quadruped; see he through two eyes as we do, or a hundred like Argus; hold he with two hands as we do, or a hundred like Briarius; walk he with two feet as we do, or a hundred like the centipede, "the mind's the standard of the man" everywhere. If he have but a wise head and a warm heart; if he be not shut up, Diogenes—like, within his own little tub of a world, but take an interest in the inhabitants of kindred spheres; and if he be a worshipper of the one God who made the heavens with all their glittering hosts;—then, in the highest sense, he is a *man*, to whom we would fain extend the hand of fellowship, claiming him as a brother in that universal family which is confined to no bone or blood, no colour or creed, and, so far as we can conjecture, to no world, but is co-extensive with the household of the Infinite Father, who cares for all of His children, and will ultimately blend them in the blessed bonds of an endless confraternity. Whether we or our posterity will ever become better acquainted in this life with the man in the moon is problematical; but in the ages to come, "when the manifold wisdom of God" shall be developed among "the principalities and powers in heavenly places," he may be something more than a myth or topic of amusement. He may be visible among the first who will declare every man in his own tongue wherein he was born the wonderful works of God, and he may be audible among the first who will lift their hallelujahs of undivided praise when every satellite shall be a chorister to laud the universal King. Let us, brothers of earth, by high and holy living, learn the music of eternity; and then, when the discord of "life's little day" is hushed, and we are called to join in the everlasting song, we may solve in one beatific moment the problem of the plurality of worlds, and in that solution we shall see more than we have been able to see at present of the man in the moon.

The Woman in the Moon

"O woman! lovely woman! nature made thee
To temper man; we had been brutes without you.
Angels are painted fair, to look like you:
There's in you all that we believe of heaven
Amazing brightness, purity, and truth,
Eternal joy, and everlasting love."

(Otway's *Venice Preserved*, 1682.)

It is not good that the man in the moon should be alone; therefore creative imagination has supplied him with a companion. The woman in the moon as a myth does not obtain to any extent in Europe; she is to be found chiefly in Polynesia, and among the native races of North America. The *Middle Kingdom* furnishes the following allusion: "The universal legend of the man in the moon takes in China a form that is at least as interesting as the ruder legends of more barbarous people. The 'Goddess of the Palace of the Moon,' Chang-o, appeals as much to our sympathies as, and rather more so than, the ancient beldame who, in European folk-lore, picks up perpetual sticks to satisfy the vengeful ideas of an ultra-Sabbatical sect. Mr. G. C. Stent has aptly seized the idea of the Chinese versifier whom he translates

"On a gold throne, whose radiating brightness
Dazzles the eyes—enhaloing the scene,
Sits a fair form, arrayed in snowy whiteness.
She is Chang-o, the beauteous Fairy Queen.
Rainbow-winged angels softly hover o'er her,
Forming a canopy above the throne;
A host of fairy beings stand before her,
Each robed in light, and girt with meteor zone.'"[62]

A touching tradition is handed down by Berthold that the moon is Mary Magdalene, and the spots her tears of repentance.[63] Fontenelle, the French poet and philosopher, saw a woman in the moon's changes. "Everything," he says, "is in perpetual motion; even including a certain young lady in the moon, who was seen with a telescope about forty years ago, everything has considerably aged. She had a pretty good face, but her cheeks are now sunken, her nose is lengthened, her forehead and chin are now prominent to such an extent, that all her charms have vanished, and I fear for her days." "What are you relating to me now?" interrupted the marchioness. "This is no jest," replied Fontenelle. "Astronomers perceived in the moon a particular figure which had the aspect of a woman's head, which came forth from between the rocks, and then occurred some changes in this region. Some pieces of mountain fell, and disclosed three points which could only serve to compose a forehead, a nose, and an old woman's chin."[64] Doubtless the face and the disfigurements were fictions of the author's lively imagination, and his words savour less of science than of satire; but Fontenelle was neither the first nor the last of those to whom "the inconstant moon that monthly changes" has been an impersonation of the fickle and the feminine. The following illustration is from Plutarch: "Cleobulus said, As touching fooles, I will tell you a tale which I heard my mother once relate unto a brother of mine. The time was (quoth she) that the moone praied her mother to make her a peticoate fit and proportionate for her body. Why, how is it possible (quoth her mother) that I should knit or weave one to fit well about thee considering that I see thee one while full, another while croissant or in the wane and pointed with tips of horns, and sometime again halfe rounde?"[65] Old John Lilly, one of our sixteenth-century dramatists, likewise supports this ungallant theory. In the *Prologus* to one of his very rare dramas he writes:

"Our poet slumb'ring in the muses laps,
Hath seen a woman seated in the moone."[66]

This woman is Pandora, the mischief-maker among the Utopian shepherds. In Act v. she receives her commission to conform the moon to her own mutability:

> "Now rule *Pandora* in fayre *Cynthia's* steede,
> And make the moone inconstant like thyselfe,
> Raigne thou at women's nuptials, and their birth,
> Let them be mutable in all their loves.
> Fantasticall, childish, and folish, in their desires
> Demanding toyes; and stark madde
> When they cannot have their will."

In North America the woman in the moon is a cosmological myth. Take, for example, the tale told by the Esquimaux, which word is the French form of the Algonquin Indian *Eskimantsic*, "raw-flesh eaters." "Their tradition of the formation of the sun and moon is, that not long after the world was formed, a great conjuror or angikak became so powerful that he could ascend into the heavens when he pleased, and on one occasion took with him a beautiful sister whom he loved very much, and also some fire, to which he added great quantities of fuel, and thus formed the sun. For a time the conjuror treated his sister with great kindness, and they lived happily together; but at last he became cruel, ill-used her in many ways, and, as a climax, burnt one side of her face with fire. After this last indignity she ran away from him and became the moon. Her brother in the sun has been in chase of her ever since; but although he sometimes gets near, will never overtake her. When new moon, the burnt side of her face is towards the earth; when full moon, the reverse is the case."[67] The likeness between this tradition and the Greenlanders' myth of Malina and Anninga is very close, the difference consisting chiefly in the change of sex; here the moon is feminine, there the moon is masculine.[68]

In Brazil the story is further varied, in that it is the sister who falls in love, and receives a discoloured face for her offence. Professor Hartt says that Dr. Silva de Coutinho found on the Rio

Branco and Sr. Barbosa has reported from the Jamundá a myth "in which the moon is represented as a maiden who fell in love with her brother and visited him at night, but who was finally betrayed by his passing his blackened hand over her face."[69]

The Ottawa tale of Indian cosmogony, called Iosco, narrates the adventures of two Indians who "found themselves in a beautiful country, lighted by the moon, which shed around a mild and pleasant light. They could see the moon approaching as if it were from behind a hill. They advanced, and the aged woman spoke to them; she had a white face and pleasing air, and looked rather old, though she spoke to them very kindly. They knew from her first appearance that she was the moon. She asked them several questions. She informed them that they were halfway to her brother's (the sun), and that from the earth to her abode was half the distance."[70]

Other American Indians have a tradition of an old woman who lived with her grand-daughter, the most beautiful girl that ever was seen in the country. Coming of age, she wondered that only herself and her grandmother were in the world. The grandam explained that an evil spirit had destroyed all others; but that she by her power had preserved herself and her grand-daughter. This did not satisfy the young girl, who thought that surely some survivors might be found. She accordingly travelled in search, till on the tenth day she found a lodge inhabited by eleven brothers, who were hunters. The eleventh took her to wife, and died after a son was born. The widow then wedded each of the others, beginning with the youngest. When she took the eldest, she soon grew tired of him, and fled away by the western portal of the hunter's lodge. Tearing up one of the stakes which supported the door, she disappeared in the earth with her little dog. Soon all trace of the fugitive was lost. Then she emerged from the earth in the east, where she met an old man fishing in the sea. This person was he who made the earth. He bade her pass into the air toward the west. Meanwhile the deserted husband pursued his wife into the earth on the west, and out again on the east, where the tantalizing old fisherman cried out to him,

"Go, go; you will run after your wife as long as the earth lasts without ever overtaking her, and the nations who will one day be upon the earth will call you *Gizhigooke*, he who makes the day." From this is derived *Gizis*, the sun. Some of the Indians count only eleven moons, which represent the eleven brothers, dying one after another.[71]

Passing on to Polynesia, we reach Samoa, where "we are told that the moon came down one evening, and picked up a woman, called Sina, and her child. It was during a time of famine. She was working in the evening twilight, beating out some bark with which to make native cloth. The moon was just rising, and it reminded her of a great bread-fruit. Looking up to it, she said, 'Why cannot you come down and let my child have a bit of you?' The moon was indignant at the idea of being eaten, came down forthwith, and took her up, child, board, mallet, and all. The popular superstition is not yet forgotten in Samoa of the *woman* in the moon. 'Yonder is Sina,' they say, 'and her child, and her mallet, and board.'"[72] The same belief is held in the adjacent Tonga group, or Friendly Islands, as they were named by Captain Cook, on account of the supposed friendliness of the natives. "As to the spots in the moon, they are compared to the figure of a woman sitting down and beating *gnatoo*" (bark used for clothing).[73]

In Mangaia, the southernmost island of the Hervey cluster, the woman in the moon is Ina, the pattern wife, who is always busy, and indefatigable in the preparation of resplendent cloth, *i.e. white clouds*. At Atiu it is said that Ina took to her celestial abode a mortal husband, whom, after many happy years, she sent back to the earth on a beautiful rainbow, lest her fair home should be defiled by death.[74] Professor Max Müller is reminded by this story of Selênê and Endymion, of Eos and Tithonos.

The Hare in the Moon

When the moon is waxing, from about the eighth day to the full, it requires no very vivid imagination to descry on the westward side of the lunar disk a large patch very strikingly resembling a rabbit or hare. The oriental noticing this figure, his poetical fancy developed the myth-making faculty, which in process of time elaborated the legend of the hare in the moon, which has left its marks in every quarter of the globe. In Asia it is indigenous, and is an article of religious belief. "To the common people in India the spots look like a hare, *i.e.* Chandras, the god of the moon, carries a hare (sasa), hence the moon is called Sasin or Sasanka, hare mark or spot."[75] Max Müller also writes, "As a curious coincidence it may be mentioned that in Sanskrit the moon is called Sasānka,*i.e.* 'having the marks of a hare,' the black marks in the moon being taken for the likeness of the hare."[76] This allusion to the sacred language of the Hindus affords a convenient opportunity of introducing one of the most beautiful legends of the East. It is a Buddhist tract; but in the lesson which it embodies it will compare very favourably with many a tract more ostensibly Christian.

"In former days, a hare, a monkey, a coot, and a fox, became hermits, and lived in a wilderness together, after having sworn not to kill any living thing. The god Sakkria having seen this through his divine power, thought to try their faith, and accordingly took upon him the form of a brahmin, and appearing before the monkey begged of him alms, who immediately brought to him a bunch of mangoes, and presented it to him. The pretended brahmin, having left the monkey, went to the coot and made the same request, who presented him a row of fish which he had just found on the bank of a river, evidently forgotten by a fisherman. The brahmin then went to the fox, who immediately went in search of food, and soon returned with a pot of milk and a dried liguan, which he had found in a plain, where apparently they had been left by a herdsman. The brahmin at last went to the hare and begged alms of him. The hare said, 'Friend, I eat nothing but grass, which I think is of no use to you.' Then the pretended brahmin replied, 'Why, friend, if you are a true hermit, you can give me your own flesh in hope of future happiness.' The hare directly consented to it, and said to the supposed brahmin, 'I have granted your request, and you may do whatever you please with me.' The brahmin then replied, 'Since you are willing to grant my request, I will kindle a fire at the foot of the rock, from which you may jump into the fire, which will save me the trouble of killing you and dressing your flesh.' The hare readily agreed to it, and jumped from the top of the rock into the fire which the supposed brahmin had kindled; but before he reached the fire, it was extinguished; and the brahmin appearing in his natural shape of the god Sakkria, took the hare in his arms and immediately drew its figure in the moon, in order that every living thing of every part of the world might see it."[77] All will acknowledge that this is a very beautiful allegory. How many in England, as well as in Ceylon, are described by the monkey, the coot, and the fox—willing to bring their God any oblation which costs them nothing; but how few are like the hare—ready to present themselves as a living sacrifice, to be consumed as a burnt offering in the Divine service! Those, however, who lose their lives in such self-sacrifice, shall find them, and be caught up to "shine as the brightness of the firmament and as the stars for ever and ever."

Another version of this legend is slightly variant. Grimm says: "The people of Ceylon relate as follows: While Buddha the great god sojourned upon earth as a hermit, he one day lost his way in a wood. He had wandered long, when a *hare* accosted him: 'Cannot I help thee? Strike into the path on thy right. I will guide thee out of the wilderness.' Buddha replied: 'Thank thee, but I am poor and hungry, and unable to repay thy kindness.' 'If thou art hungry,' said the hare, 'light a fire, and kill, roast, and eat me.' Buddha made a fire, and the hare immediately jumped in. Then did Buddha manifest his divine power; he snatched the beast out of the flames, and set him in the moon, where he may be seen to this day."[78] Francis Douce, the antiquary, relates this myth, and adds, "this is from the information of a learned and intelligent French gentleman recently arrived from Ceylon, who adds that the Cingalese would often request of him to permit them to look for the hare through his telescope, and exclaim in raptures that they saw it. It is remarkable that the Chinese represent the moon by a rabbit pounding rice in a mortar. Their mythological moon Jut-ho is figured by a beautiful young woman with a double sphere behind her head, and a rabbit at her feet. The period of this animal's gestation is thirty days; may it not therefore typify the moon's revolution round the earth."[79]

Sâkyamuni as a Hare in the Moon
Collin de Plancy's "Dictionnaire Infernal."

In this same apologue we have doubtless a duplicate, the original or a copy, of another Buddhist legend found among the Kalmucks of Tartary; in which Sâkyamuni himself, in an early stage of existence, had inhabited the body of a hare. Giving himself as food to feed the hunger of a starving creature, he was immediately placed in the moon, where he is still to be seen.[80]

The Mongolian also sees a hare in the lunar shadows. We are

told by a Chinese scholar that "tradition earlier than the period of the Han dynasty asserted that a hare inhabited the surface of the moon, and later Taoist fable depicted this animal, called the gemmeous hare, as the servitor of the genii, who employ it in pounding the drugs which compose the elixir of life. The connection established in Chinese legend between the hare and the moon is probably traceable to an Indian original. In Sanskrit inscriptions the moon is called Sason, from a fancied resemblance of its spots to a leveret; and pandits, to whom maps of the moon's service have been shown, have fixed on *Loca Paludosa*, and *Mons Porphyrites* or *Keplerus* and *Aristarchus*, for the spots which they think exhibit the similitude of a hare."[81] On another page of the same work we read: "During the T'ang dynasty it was recounted that a cassia tree grows in the moon, this notion being derived apparently from an Indian source. The *sal* tree (*shorea robusta*), one of the sacred trees of the Buddhists, was said during the Sung dynasty to be identical with the cassia tree in the moon. The lunar hare is said to squat at the foot of the cassia tree, pounding its drugs for the genii. The cassia tree in the moon is said to be especially visible at mid-autumn, and hence to take a degree at the examinations which are held at this period is described as plucking a leaf from the cassia."[82]

This hare myth, attended with the usual transformation, has travelled to the Hottentots of South Africa. The fable which follows is entitled "From an original manuscript in English, by Mr. John Priestly, in Sir G. Grey's library." "The moon, on one occasion, sent the hare to the earth to inform men that as she (the moon) died away and rose again, so mankind should die and rise again. Instead, however, of delivering this message as given, the hare, either out of forgetfulness or malice, told mankind that as the moon rose and died away, so man should die and rise no more. The hare, having returned to the moon, was questioned as to the message delivered, and the moon, having heard the true state of the case, became so enraged with him that she took up a hatchet to split his head; falling short, however, of that, the hatchet fell upon the upper lip of the hare, and cut it severely. Hence it is that we see the 'hare-lip.' The hare, being duly incensed at having received such treatment, raised his claws, and scratched the moon's face; and the dark parts which we now see on the surface of the moon are the scars which she received on that occasion."[83] In an account of the Hottentot myth of the "Origin of Death," the angered moon heats a stone and burns the hare's mouth, causing the hare-lip.[84] Dr. Marshall may tell us, with all the authority of an eminent physiologist, that hare-lip is occasioned by an arrest in the development of certain frontal and nasal processes,[85] and we may receive his explanation as a sweetly simple solution of the question; but who that suffers from this leporine-labial deformity would not prefer a supernatural to a natural cause? Better far that the lip should be cleft by Shakespeare's "foul fiend Flibbertigibbet," than that an abnormal condition should be accounted for by science, or comprised within the reign of physical law.

Even Europe is somewhat hare-brained: for Caesar tells us that the Britons did not regard it lawful to eat the hare, though he does not say why; and in Swabia still, children are forbidden to make shadows on the wall to represent the sacred hare of the moon.

We may pursue this matter even in Mexico, whose deities and myths a recent Hibbert lecturer brought into clearer light, showing that the Mexicans "possessed beliefs, institutions, and a developed mythology which would bear comparison with anything known to antiquity in the old world."[86] The Tezcucans, as they are usually called, are described by Prescott as "a nation of the same great family with the Aztecs, whom they rivalled in power, and surpassed in intellectual culture and the arts of social refinement."[87] Their account of the creation is that "the sun and moon came out equally bright, but this not seeming good to the gods, one of them took a rabbit by the heels and slung it into the face of the moon, dimming its lustre with a blotch, whose mark may be seen to this day."[88]

We have now seen that the fancy of a hare in the moon is universal; but not so much importance is to be attached to this, as to some other aspects of moon mythology. The hare-like patch is visible in every land, and suggested the animal to all observers. That the rabbit's period of gestation is thirty days is a singular coincidence; but that is all—nay, it is not even that, for "the moon's revolution round the earth," which Douce supposed the Chinese myth to typify, is accomplished in a little more than *twenty-seven* days. Neither is much weight due to the fanciful comparison of Gubernatis: "The moon is the watcher of the sky, that is to say, she sleeps with her eyes open; so also does the hare, whence the *somnus leporinus* became a proverb."[89] The same author says on another page, and here we follow him: "The mythical hare is undoubtedly the moon. In the first story of the third book of the *Pancatantram*, the hares dwell upon the shore of the lake Candrasaras, or lake of the moon, and their king has for his palace the lunar disk."[90] It is this story, which Mr. Baring-Gould relates in outline; and which we are compelled still further to condense. In a certain forest there once lived a herd of elephants. Long drought having dried up the lakes and swamps, an exploring party was sent out in search of a fresh supply of water. An extensive lake was discovered, called the moon lake. The elephants with their king eagerly marched to the spot, and found their thirsty hopes fully realized. All round the lake were in numerable hare warrens, which the tread of the mighty monsters crushed unmercifully, maiming and mangling the helpless inhabitants. When the elephants had withdrawn, the poor hares met together in terrible plight, to consult upon the course which they should take when their enemies returned. One wise hare undertook the task of driving the ponderous herd away. This he did by going alone to the elephant king, and representing himself as the hare which lived in the moon. He stated that he was deputed by his excellency the moon to say that if the elephants came any more to the lake, the beams of night would be withheld, and their bodies would be burned up with perpetual sunshine. The king of the elephants thinking that "the better part of valour is discretion," decided to offer an apology for his offence. He was conducted to the lake, where the moon was reflected in the water, apparently meditating his revenge. The elephant thrust his proboscis into the lake, which disturbed the reflection. Whereupon the elephant, judging the moon to be enraged, hurried with his apology, and then went off vowing never to return. The wise hare had proven that "wisdom is better than strength"; and the hares suffered no more molestation. "We may also remark, in this event, the truth of that saying of Euripides, 'that one wise counsel is better than the strength of many'" (*Polybius*, i. 35).

The Toad in the Moon

We owe an immense debt of gratitude and honour to the many enterprising and cultivated men who have gone into all parts of the earth and among all peoples to investigate human history and habit, mythology and religion, and thus enrich the stores of our national literature. With such a host of travellers gathering up the fragments, nothing of value is likely to be lost. We have to thank intelligent explorers for all we know of the mythical frog or toad in the moon: an addition to our information which is not unworthy of thoughtful notice.

The Selish race of North-west American Indians, who inhabit the country between the Cascade and Rocky Mountains, have a tradition, which Captain Wilson relates as follows: "The expression of 'a toad in the moon,' equivalent to our 'man in the moon,' is explained by a very pretty story relating how the little wolf, being desperately in love with the toad, went a-wooing one night and prayed that the moon might shine brightly on his adventure; his prayer was granted, and by the clear light of a full moon he was pursuing the toad, and had nearly caught her, when, as a last chance of escape, she made a desperate spring on to the face of the moon, where she remains to this day."[91] Another writer says that "the Cowichan tribes think that the moon has a frog in it."[92]

From the Great Western we turn to the Great Eastern world, and in China find the frog in the moon. "The famous astronomer Chang Hêng was avowedly a disciple of Indian teachers. The statement given by Chang Hêng is to the effect that 'How I, the fabled inventor of arrows in the days of Yao and Shun,* obtained the drug of immortality from Si Wang Mu (the fairy 'Royal Mother' of the West); and Chang Ngo (his wife) having stolen it,

* Mr. Herbert A. Giles says that How I was a legendary chieftain, who "flourished about 2,500 B.C." *Strange Stories from a Chinese Studio*, London, 1880, i. 19, *note*.

fled to the moon, and became the frog—*Chang-chu*—which is seen there.' The lady *Chang-ngo* is still pointed out among the shadows in the surface of the Moon."[93] Dr. Wells Williams also tells us that in China "the sun is symbolized by the figure of a raven in a circle, and the moon by a rabbit on his hind legs pounding rice in a mortar, or by a three-legged toad. The last refers to the legend of an ancient beauty, Chang-ngo, who drank the liquor of immortality, and straightway ascended to the moon, where she was transformed into a toad, still to be traced in its face. It is a special object of worship in autumn, and moon cakes dedicated to it are sold at this season."[94] We have little doubt that what the Chinese look for they see. We in the West characterize and colour objects which we behold, as we see them through the painted windows of our predisposition or prejudice. As a great novelist writes: "From the same object different conclusions are drawn; the most common externals of nature, the wind and the wave, the stars and the heavens, the very earth on which we tread, never excite in different bosoms the same ideas; and it is from our own hearts, and not from an outward source, that we draw the hues which colour the web of our existence. It is true, answered Clarence. You remember that in two specks of the moon the enamoured maiden perceived two unfortunate lovers, while the ambitious curate conjectured that they were the spires of a cathedral."[95] Besides, it must be confessed that the particular moon-patch that has awakened so much interest in every age and nation is quite as much like a frog or toad as it is like a rabbit or hare.

Other Moon Myths

It is almost time that we should leave this lunar zoology; we will therefore merely present a few creatures which may be of service in a comparative anatomy of the whole subject, and then close the account. There is a story told in the Fiji Islands which so nearly approaches the Hottentot legend of the hare, that they both seem but variations of a common original. In the one case the opponent of the moon's benevolent purpose affecting man's hereafter was a hare, in the other a rat. The story thus runs: There was "a contest between two gods as to how man should die. Ra Vula (the moon) contended that man should be like himself—disappear awhile, and then live again. Ra Kalavo (the rat) would not listen to this kind proposal, but said, 'Let man die as a rat dies.' And he prevailed."[96] Mr. Tylor, who quotes this rat story, adds: "The dates of the versions seem to show that the presence of these myths among the Hottentots and Fijians, at the two opposite sides of the globe, is at any rate not due to transmission in modern times."[97]

From the rat to one of its mortal enemies is an easy transition. The Australian story is that Mityan, the moon, was a native cat, who fell in love with another's wife, and while trying to induce her to run away with him, was discovered by the husband, when a fight took place. Mityan was beaten and ran away, and has been wandering ever since.[98] We are indebted for another suggestion to Bishop Wilkins, who wrote over two centuries ago: "As for the form of those spots, *Albertus* thinks that it represents a lion, with his tail towards the east, and his head the west; and some others have thought it to be very much like a fox, and certainly 'tis as much like a lion as that in the *zodiac*, or as *ursa major* is like a bear."[99] This last remark of the old mathematician is "a hit, a very palpable hit," at those unpoetical people who catalogue the constellations under all sorts of living creatures' names, implying resemblances, and then "sap with solemn sneer" our myths of the moon.

We have now seen that the moon is populated with men, women, and children,—hares and rabbits, toads and frogs, cats and dogs, and sundry small "cattle"; we observe in making our exit that it is also planted with a variety of trees; in short, is a zoological garden of a high order. Even among the ancients some said the lunar spots were forests where Diana hunted, and that the bright patches were plains. Captain Cook tells us that in the South Pacific "the spots observed in the moon are supposed to be groves of a sort of trees which once grew in Otaheite, and, being destroyed by some accident, their seeds were carried up thither by doves, where they now flourish."[100] Ellis also tells of these Tahitians that "their ideas of the moon, which they called *avae* or *marama*, were as fabulous as those they entertained of the sun. Some supposed the moon was the wife of the sun; others that it was a beautiful country in which the aoa grew."[101] These arborary fancies derive additional interest, if not a species of verisimilitude, from the record of a missionary that "a stately tree, clothed with dark shining leaves, and loaded with many hundreds of large green or yellowish-coloured fruit, is one of the most splendid and beautiful objects to be met with among the rich and diversified scenery of a Tahitian landscape."

Our collection of lunar legends is now on exhibition. No thoughtful person will be likely to dispute the dictum of Sir John Lubbock that "traditions and myths are of great importance, and indirectly throw much light on the condition of man in ancient times."[102] But they serve far more purposes than this. They are the raw material, out of which many of our goodly garments of modern science and religion are made up. The illiterate negroes on the cotton plantation, and the rude hunters in the jungle or seal fishery, produce the staple, or procure the skins, which after long labour afford comfort and adornment to proud philosophers and peers. The golden cross on the saintly bosom and the glittering crown on the sovereign brow were embedded as rough ore in primeval rocks ages before their wearers were born to boast of them. We shall esteem our treasures none the less because their origin is known, as we love "the Best of men" none the less because he was born of a woman. We closed our series of moon myths with a vision of a beautiful country, ornamented with groves of fruitful trees, whose seeds had been carried thither by white-winged doves; and carried thither because "some accident" had destroyed the trees in their native isles on earth. Thus the lunar world had become a desirable scene of superior and surpassing loveliness. Who can reflect upon this dream of human childhood, and not recall some dreams of later years? Who can fail to discern slight touches of the same hand which we see displayed in other designs? "Happily for historic truth," says Mr. Tylor, "mythic tradition tells its tales without expurgating the episodes which betray its real character to more critical observation."[103] Who is not led on from Tahiti to Greece, and to the Isles of the Blessed, the Elysium which abounds in every charm of life, and to the garden of the Hesperides, with its apples of gold; thence to the Meru of the Hindoos, the sacred mountain which is perpetually clothed in the rays of the sun, and adorned with every variety of plants and trees; thence again to the Heden of the Persians, of matchless beauty, where ever flourishes the tree Hom with its wonderful fruit; on to the Chinese garden, near the gate of heaven, whose noblest spring is the fountain of life, and whose delightful trees bear fruits which preserve and prolong the existence of man?[104] Thence an easy entrance is gained to the Hebrew Paradise, with its abounding trees "pleasant to the sight and good for food, the tree of life also in the midst of the garden"; and finally arises a sight of the "better land" of the Christian poetess, the incorruptible and undefiled inheritance of the Christian preacher, the prospect which is "ever vernal and blooming,—and, best of all, amid those trees of life there lurks no serpent to destroy,—the country, through whose vast region we shall traverse with untired footsteps, while every fresh revelation of beauty will augment our knowledge, and holiness, and joy."[105] Who will travel on such a pilgrimage of enlarged thought, and not come to the conclusion that if one course of development has been followed by all scientific and spiritual truth, then "almost the whole of the mythology and theology of civilized nations maybe traced, without arrangement or co-ordination, and in forms that are undeveloped and original rather than degenerate, in the traditions and ideas of savages"?[106] Such a conclusion may diminish our self-esteem, if we have supposed ourselves the sole depositaries of Divine knowledge; but it will exalt our conception of the generosity of the Father of all men, who never left a human soul without a witness of His invisible presence and ineffable love.

MOON WORSHIP

Introduction

We have now to show that the moon has been in every age, and remains still, one of the principal objects of human worship. Even among certain nations credited with pure monotheism, it will be manifested that there was the practice of that primitive polytheism which adored the hosts of heaven. And, however humiliating or disappointing the disclosure may prove, it will be established that some of the foremost Christian peoples of the world maintain luniolatry to this day, notwithstanding that they have the reproving light of the latest civilization. We are so prone to talk of heathenism as abroad, that we forget or neglect the gross heathenism which abounds at home; and while we complacently speak of the march of the world's progress with which we identify ourselves, we are oblivious of the fact that much ancient falsehood survives and blends with the truth in which our superior minds, or minds with superior facilities, have been trained. How few of us reflect that the signs and symbols of rejected theories have passed into the nomenclature of received systems! Nay, we plume ourselves upon the new translation or revision as if we were the favoured recipients of some fresh revelation. Not only in the names of our days and months, but also in some of our most cherished dogmas, we are but the "liberal-conservatives" in religion, who retain the old, while we congratulate ourselves upon being the apostles of the new. That the past must always run into the present, and the present proceed from the past, we readily enough allow as a natural and necessary law; yet baptized heathenism is often heathenism still, under another name. Again, we are sometimes so short-sighted that we deny to former periods the paternity of their own more fortunate offspring, and behave like prosperous children who ungratefully ignore their poorer parents, to whom they owe their breath and being. Such treatment of history is to be emphatically deprecated, whether it arises from ignorance or ingratitude. We ought to know, if we do not, and we ought also to acknowledge, that our perfect day grew out of primeval darkness, and that the progress was a lingering dawn. This we hold to be the clearest view of the Divine causation. Our modern method in philosophy, largely owing to the *Novum Organum* of Bacon, is evolution, the *novum organum* of the nineteenth century; and this process recognises no abrupt or interruptive creations, but gradual transformations from pre-existent types, "variations under domestication," and the passing away of the old by its absorption into the new. Our religion, like our language, is a garden not only for indigenous vegetation, but also for acclimatisation, in which we improve under cultivation exotic plants whose roots are drawn from every soil on the earth. And, as Paul preached in Athens the God whom the Greeks worshipped in ignorance, so our missionaries carry back to less enlightened peoples the fruit of that life-giving tree whose germs exist among themselves, undeveloped and often unknown. No religion has fallen from heaven, like the fabled image of Athene, in full-grown beauty. All spiritual life is primordially an inspiration or intuition from the Father of spirits, whose offspring all men are, and who is not far from every one of them. This intuition prompts men to "seek the Lord, if haply they might feel after Him, and find Him." Thus prayer becomes an instinct; and to worship is as natural as to breathe. But man is a being with five senses, and as his contact with his fellow-creatures and with the whole creation is at one or other of those five points, he is necessarily sensuous. Endowed with native intelligence, the *intellectus ipse* of Leibnitz, he nevertheless receives his impressions on *sensitive* nerves, his emotions are *sentiments*, his words become *sentences*, and his stock of wisdom is his common *sense*. A few, very few, words express his sensations, a few more his perceptions, and so on; but he is conscious of *objects* at first, he deals with *subjects* afterwards. Soon the sun, moon, and stars, as bright lights attract his eyes, as we have all seen an infant of a few days fix its gaze upon a candle or lamp. These heavenly orbs are found to be in motion, to be far away, to be the glory of day and night: what wonder if *ideas* of these *images* are formed in the religious mind, if the worshipper imagines the sun and moon to be reflections of the God of light, and pays homage to the creature which renders the Creator visible? Thus in the childhood of man religion grows, and with the multiplication of intellect and sensation, endless diversity of language, conception and faith is the result. Another result, of course, is the endless diversity of deities. Every race, every nation, every tribe, every household, every heart, has had its own God. And yet, with all this multiplicity in religious literature and dogma, subject and object, a unity co-exists which the student of the science notes with profound interest. All nations of men are of one blood; and all forms of God embody the one Eternal Spirit. To this unity mythology tends. As one writer says: "We must ever bear in mind that the course of mythology is from many gods toward one, that it is a synthesis, not an analysis, and that in this process the tendency is to blend in one the traits and stories of originally separate divinities."[107] The ancient Hebrew worshipped God as "the Eternal, our righteousness"; the Greek worshipped Him as wisdom and beauty; the Roman as power and government; the Persian as light and goodness; and so forth. Few hymns have surpassed the beauty of Pope's *Universal Prayer.* It is the *Te Deum laudamus* of that catholic Church which embraces God-loved humanity.

"Father of all! in every age,
In every clime, adored,
By saint, by savage, and by sage,
Jehovah, Jove, or Lord!"

The Christian, believing his to be the "One Religion," as a recent Bampton Lecturer termed it, too often forgets that his system is a recomposition of rays of a religious light which was decomposed in the prismatic minds of earlier men. And further, with a change of metaphor, if Christianity has flourished and fructified through eighteen centuries, it must not be denied that it is a graft upon an old stock which through fifteen previous centuries had borne abundant fruit. The same course must be adopted still. We find men everywhere holding some truth; we add further truth; until, as a chemist would say, we saturate the solution, which upon evaporation produces a crystallized life of entirely new colour and quality and form. Thus Professor Nilsson writes: "Every religious *change* in a people is, in fact, only an intermixture of religions; because the new religion, whether received by means of convincing arguments, or enforced by the eloquence of fire and sword, cannot *at once* tear up all the wide-spreading roots by which its forerunner has grown in the heart of the people; this must be the work of many years, perhaps of many generations."[108] We cannot better close this lengthy introduction than by reminding Christians of the saying of their Great and Good Teacher, "I am not come to destroy, but to fulfil."

The Moon Mostly a Male Deity

We have already in part pointed out that the moon has been considered as of the masculine gender; and have therefore but to travel a little farther afield to show that in the Aryan of India, in Egyptian, Arabian, Slavonian, Latin, Lithuanian, Gothic, Teutonic, Swedish, Anglo-Saxon, and South American, the moon is a male god. To do this, in addition to former quotations, it will be sufficient to adduce a few authorities. "Moon," says Max Müller, "is a very old word. It was *móna* in Anglo-Saxon, and was used there, not as a feminine, but as a masculine for the moon was originally a masculine, and the sun a feminine, in all Teutonic languages; and it is only through the influence of classical models that in English moon has been changed into a feminine, and sun into a masculine. It was a most unlucky assertion which Mr. Harris made in his *Hermes*, that all nations ascribe to the sun a masculine, and to the moon a feminine gender."[109] Grimm says, "Down to recent times, our people were fond of calling the sun and moon *frau sonne* and *herr mond*."[110] Sir Gardner Wilkinson writes: "Another reason that the moon in the Egyptian mythology could not be related to Bubastis is, that it was a male and not a female deity, personified in the god Thoth. This was also the case in some religions of the West. The Romans recognised the god Lunus; and the Germans, like the Arabs, to this day, consider the moon masculine, and not feminine, as were the Selênê and Luna of the Greeks and Latins."[111] Again, "The Egyptians represented their moon as a male deity, like the German *mond* and *monat*, or the *Lunus* of the Latins; and it is worthy of remark, that the same custom of calling it male is retained in the East to the present day, while the sun is considered female, as in the language of the Germans."[112] "In Slavonic," Sir George Cox tells us, "as in the Teutonic mythology, the moon is male. His wedding with the sun brings on him the wrath of Perkunas[the thunder-god], as the song tells us

'The moon wedded the sun
In the first spring.
The sun rose early
The moon departed from her.
The moon wandered alone;
Courted the morning star.
Perkunas, greatly wroth,
Cleft him with a sword.
'Wherefore dost thou depart from the sun,
Wandering by night alone,
Courting the morning star?'"[113]

"In a Servian song a girl cries to the sun—

'O brilliant sun! I am fairer than thou
Than thy brother, the bright moon.'"

In South Slavonian poetry the sun often figures as a radiant youth. But among the northern Slavonians, as well as the Lithuanians, the sun was regarded as a female being, the bride of the moon. 'Thou askest me of what race, of what family I am,' says the fair maiden of a song preserved in the Tambof Government—

'My mother is—the beauteous Sun,
And my father—the bright Moon.'"[114]

"Among the Mbocobis of South America the moon is a man and the sun his wife."[115] The Ahts of North America take the same view; and we know that in Sanskrit and in Hebrew the word for moon is masculine.

This may seem to many a matter of no importance; but if mythology throws much light upon ancient history and religion, its importance may be considerable, especially as it lies at the root of that sexuality which has been the most prolific parent of both good and evil in human life. The sexual relation has existed from the very birth of animated nature; and it is remarkable that a man of learning and piety in Germany has made the strange if not absurd statement that in the beginning "Adam was externally sexless."[116] Another idea, more excusable, but equally preposterous, is, that grammatical gender has been the cause of the male and female personation of deities, when really it has been the result. The cause, no doubt, was inherent in man's constitution; and was the inevitable effect of thought and expression. The same necessity of natural language which led the Hebrew prophets to speak of their land as married, of their nation as a wife in prosperity and a widow in calamity, of their Maker as their husband, who rejoices over them as the bridegroom rejoiceth over the bride:[117] this same necessity, becoming a habit like that of our own country folks in Hampshire, of whom Cobbett speaks, who call almost everything *he* or *she*; led the sensuous and imaginative ancients, as it leads simple and poetical peoples still, to call the moon a man and to worship him as a god. Objects of fear and reverence would be usually masculines; and objects of love and desire feminines. We may thus find light thrown upon the honours paid to such goddesses as Astarte and Aphrodite: which will also help us to understand the deification by a celibate priesthood of the Virgin Mary. We may, moreover, account partly for the fact that to the sailor his ship is always she; to the swain the flowers which resemble his idol, as the lily and the rose, are always feminine, and used as female names; while to the patriot the mother country is nearly always of the tender sex. [118] Prof. Max Müller thinks that the distinction between males and females began, "not with the introduction of masculine nouns, but with the introduction of feminines, *i.e.* with the setting apart of certain derivative suffixes for females. By this all other words became masculine."[119] Thus the sexual emotions of men created that grammatical gender which has contributed so powerfully to our later mythology, and has therefore been mistaken for the author of our male and female personations. What beside sexuality suggested the thought of the Chevalier Marini? "He introduces the god *Pan*, who boasts that the spots which are seen in the moon are impressions of the kisses he gave it."[120] That grammar is very much younger than sexual relations is proven by the curious fact mentioned by Max Müller that *pater* is not a masculine, nor *mater* a feminine. Gender, we must not forget, is from *genus*, a kind or class; and that the classification in various languages has been arranged on no fixed plan. We in our modern English, with much still to do, have improved in this respect, since, in Anglo-Saxon, *wif* = wife, was neuter, and *wif-mann* = woman, was masculine. In German still *die frau*, the woman, is feminine; but *das weib*, the wife, is neuter.[121] Dr. Farrar finds the root of gender in the imagination: which we admit if associated with sex. Otherwise, we cannot understand how an *unfelt* distinction of this sort could be mentally *seen*. But Dr. Farrar means more than imagination, for he says, "from this source is derived the whole system of genders for inanimate things, which was perhaps inevitable at that early childish stage of the human intelligence, when the actively working soul attributed to everything around it some portion of its own life. Hence, well-nigh everything is spoken of as masculine or feminine."[122] We are surprised that Dr. Farrar seems to think German an exception, in making a masculine noun of the moon. He has failed to apply to this point his usual learned and laborious investigation.[123]

Diogenes Laertius describes the theology of the Jews as an offshoot from that of the Chaldees, and says that the former affirm of the latter "that they condemn images, and especially those persons who say that the gods are male and female."[124] Which condemnation implies the prevalence of this sexual distinction between their deities.

In concluding this chapter we think that it will be granted that gender in the personification of inanimate objects was the result of sex in the animate subject: that primitive men saw the moon as a most conspicuous object, whose spots at periods had the semblance of a man's face, whose waxing and waning increased their wonder: whose coming and going amid the still and solemn night added to the mystery: until from being viewed as a man, it was feared, especially when apparently angry in a mist or an eclipse, and so reverenced and worshipped as the heaven-man, the monthly god.

The Moon a World-Wide Deity

Anthropomorphism, or the representation of outward objects in the *form* of *man*, wrought largely, as we have seen, in the manufacture of the man in the moon; it entered no less into the composition of the moon-god. The twenty-first verse of the fiftieth Psalm contains its recognition and rebuke. "Thou thoughtest that I was altogether as thyself"; or, still more literally, "Thou hast thought that being, I shall be like thee." As Dr. Delitzsch says, "Because man in God's likeness has a bodily form, some have presumed to infer backwards therefrom that God also has a bodily form like to man, which is related by way of prototype to the human form."[125] As well might we say that because a watchmaker constructs a chronometer with a movement somewhat like that of his own heart, therefore he is mechanical, metallic, and round. Against this anthropomorphic materialism science lifts up its voice; for what modern philosopher, worthy of the name, fails to distinguish between phenomenon and fact, inert matter and active force? Says a recent writer, "We infer that as our own master of the mint is neither a sovereign nor a half-sovereign, so the force which coins and recoins this υλη, or matter, must be altogether in the god-part and none of it in the metal or paste in which it works."[126] With the progress of man's intelligence we shall observe improvement in this anthropomorphism, but it will still survive. As Mr. Baring-Gould tells us: "The savage invests God with bodily attributes; in a more civilized state man withdraws the bodily attributes, but imposes the limitations of his own mental nature; and in his philosophic elevation he recognises in God intelligence only, though still with anthropomorphic conditions."[127]

Xenophanes said that if horses, oxen, and lions could paint, they would make gods like themselves. And Ralph Waldo Emerson says: "The gods of fable are the shining moments of great men. We run all our vessels into one mould. Our colossal theologies of Judaism, Christism, Buddhism, Mahometism, are the necessary and structural action of the human mind. The student of history is like a man going into a warehouse to buy clothes or carpets. He fancies he has a new article. If he go to the factory, he shall find that his new stuff still repeats the scrolls and rosettes which are found on the interior walls of the pyramids of Thebes. Our theism is the purification of the human mind. Man can paint, or make, or think nothing but man. He believes that the great material elements had their origin from his thought. And our philosophy finds one essence collected or distributed."[128] And a devout author, whose orthodoxy—whatever that may mean—is unquestioned, acknowledges that man adored the unknown power in the sun, and "in the moon, which bathes the night with its serene splendours. Under this latter form, completed by a very simple anthropomorphism which applies to the gods the law of the sexes, the religions of nature weighed during long ages upon Western Asia."[129] A volume might be written upon this subject; but we have other work in hand.

It seems to be generally admitted that no form of idolatry is older than the worship of the moon. Lord Kames says, "It is probable that the sun and moon were early held to be deities, and that they were the first visible objects of worship."[130] Dr. Inman says, "That the sun and moon were at a very early period worshipped, none who has studied antiquity can deny."[131] And Goldziher maintains that "the lunar worship is older than the solar."[132] Maimonides, "the light of Israel," says that the Zabaists not only worshipped the moon themselves, but they also asserted that Adam led mankind to that species of worship. No doubt luniolatry is as old as the human race. In some parts the moon is still the superior god. Mr. Tylor writes: "Moon worship, naturally ranking below sun worship in importance, ranges through nearly the same district of culture. There are remarkable cases in which the moon is recognised as a great deity by tribes who take less account, or none at all, of the sun. An old account of the Caribs describes them as esteeming the moon more than the sun, and at new moon coming out of their houses crying, Behold the moon!"[133] This deity, then, is ancient and modern: also a chief of the gods: let us now show that he is a god whose empire is the world.

We begin in Asia, and with the Assyrian monuments, which display many religious types and emblems. "Representations of the heavenly bodies, as sacred symbols, are of constant occurrence in the most ancient sculptures. In the bas-reliefs we find figures of the sun, moon, and stars, suspended round the neck of the king when engaged in the performance of religious ceremonies."[134] In Chaldaea "the moon was named Sin and Hur. Hurki, Hur, and Ur was the chief place of his worship, for the satellite was then considered as being masculine. The name for the moon in Armenian was *Khaldi*, which has been considered by some to be the origin of the word Chaldee, as signifying moon worshippers."[135] With this Chaldaean deity may be connected "the Akkadian moon god, who corresponds with the Semitic Sin," and who "is Aku, 'the seated-father,' as chief supporter of kosmic order, styled 'the maker of brightness,' En-zuna, 'the lord of growth,' and Idu, 'the measuring lord,' the Aïdês of Hesychios."[136]

"With respect to the name of Chaldaean, perhaps the most probable account of the origin of the word is, that it designates properly the inhabitants of the ancient capital, Ur or Hur,—*Kkaldi* being in the Burbur dialect the exact equivalent of *Hur*, which was the proper name of the moon god, and Chaldaeans being thus either 'moon worshippers,' or simply, inhabitants of the town dedicated to, and called after, the moon."[137] Again: "The first god of the second triad is Sin or Hurki, the moon deity. It is in condescension to Greek notions that Berosus inverts the true Chaldaean order, and places the sun before the moon in his enumeration of the heavenly bodies. Chaldaean mythology gives a very decided preference to the lesser luminary, perhaps because the nights are more pleasant than the clays in hot countries. With respect to the names of the god, we may observe that Sin, the Assyrian or Semitic term, is a word of quite uncertain etymology, which, however, is found applied to the moon in many Semitic languages."[138] "*Sin* is used for the moon in Mendaean and Syriac at the present day. It is the name given to the moon god in St. James of Seruj's list of the idols of Harran; and it was the term used for Monday by the Sabaeans as late as the ninth century."[139] Another author writes: "The Babylonian and Assyrian moon god is Sin, whose name probably appears in Sinai. The expression, 'from the origin of the god Sin,' was used by the Assyrians to mark remote antiquity; because, as chaos preceded order, so night preceded day, and the enthronement of the moon as the night-king marks the commencement of the annals of kosmic order."[140]

When we search the Hebrew Scriptures, we find too many allusions to the Queen of Heaven, to Astarte and the groves, for us to doubt that the Israelites adored

"—moonèd Ashtaroth,
Heaven's queen and mother both." (Milton's *Odes*.)

Dr. Goldziher is an incontestable authority, and thus writes: "Queen or Princess of Heaven is a very frequent name for the moon."[141] Again, "Even in the latest times the Hebrews called the moon the 'Queen of Heaven' (Jer. vii. 18), and paid her Divine honours in this character at the time of the captivity."[142] And, to complete this author's witness, he again says: "What was the antiquity of this lunar worship among the Hebrews, is testified (as has long been known) by the part played by Mount Sinai in the history of Hebrew religion. For this geographical name is doubtless related to *Sin*, one of the Semitic names of the moon. The mountain must in ancient times have been consecrated to the moon. The beginning of the Hebrew religion, which was connected with the phenomena of the night-sky, germinated first during the residence in Egypt on the foundation of an ancient myth. The recollection of this occasioned them to call the part of Egypt which they had long inhabited, eres Sînîm, 'moonland' (Isa. xlix. 12)."[143] It is but just that we should hear the other side, when there is a difference of opinion. The above mentioned 'Queen of Heaven' is beyond question the Ashtoreth or Astarte (identical with our *star*), which was the principal goddess of the Phoenicians; and we believe she was originally the goddess of the moon. This is doubted by a modern writer, who says, "Baal is constantly coupled with Astarte; and the more philosophical opinion is that this national god and goddess were the lord and lady of Phoenicia, rather than the sun and moon: for to a people full of political life the sun and moon would have been themselves representatives, while a Divine king and queen were the realities. And if so, the habitual inclination of the Israelites, an essentially political people, for this worship becomes the more easily understood."[144] Professor F. D. Maurice, in his *Moral and Metaphysical Philosophy*, also takes this view. The question here is not whether the Jews worshipped Astarte, but whether Astarte was the moon. This we cannot hesitate to answer in the affirmative. Kenrick writes: "Ashtoreth or Astarte appears physically to represent the moon. She was the chief local deity of Sidon; but her worship must have been extensively diffused, not only in Palestine, but in the countries east of the Jordan, as we find Ashtaroth-Karnaim (Ashtaroth of two horns) mentioned in the book of Genesis (xiv. 5). This goddess, like other lunar deities, appears to have been symbolized by a heifer, or a figure with a heifer's head, whose horns resembled the crescent moon. The children of Israel renounced her worship at the persuasion of Samuel; and we do not read again of her idolatry till the reign of Solomon (1 Kings xi. 5), after which it appears never to have been permanently banished, though put down for a time by Josiah (2 Kings xxiii. 13). She is the Queen of Heaven, to whom, according to the reproaches of Jeremiah (vii. 18, xliv. 25), the women of Israel poured out their drink-offerings, and burnt incense, and offered cakes, regarding her as the author of their national prosperity. This epithet accords well with the supposition that she represented the moon, as some ancient authors inform us."[145] Dr. Gotch, an eminent Hebrew scholar, says that there is no doubt that the moon is the symbol of productive power and must be identified with Astarte. "That this goddess was so typified can scarcely be doubted. The ancient name of the city, Ashtaroth-Karnaim, already referred to, seems to indicate a horned Astarte, that is an image with a crescent moon on her head like the Egyptian Athor. At any rate, it is certain that she was by some ancient writers identified with the moon, as Lucian and Herodian. On these grounds Movers, Winer, Keil, and others maintain that originally Ashtoreth was the moon goddess."[146] Clearly, then, the Hebrews worshipped the moon. But, even apart from Astarte, this worship may be proven on other evidence. Dr. Jamieson says that the word *mena* (moon: Anglo-Saxon, *mona*) "approaches most nearly to a word used by the prophet Isaiah, which has been understood by the most learned interpreters as denoting the moon. 'Ye are they that prepare a table for *Gad*, and that furnish the offering unto *Meni*.' (Isa. lxv. 11). As *Gad* is understood of the *sun*, we learn from Diodor Sicul that *Meni* is to be viewed as a designation of the *moon*."[147] This is Bishop Lowth's view. "The disquisitions and conjectures of the learned concerning Gad and Meni are infinite and uncertain: perhaps the most probable may be, that Gad means good fortune, and Meni the moon."[148] One point is worthy of notice. In our English version *Meni* is rendered "number"; and we know very well that by the courses of the moon ancient months and years were numbered. In Isaiah iii. 18 we find the daughters of Zion ornamented with feet-rings, and networks, and *crescents*: or, as our translation reads, "round tires like the moon." And, once more, in Ezekiel xlvi., we read that the gate of the inner court of the sanctuary that "looketh toward the east, shall be opened on the day of the new moon"; and the meat offering on "the day of the new moon shall be a young bullock without blemish, and six lambs, and a ram." If there was no sacred significance in the observance of these lunar changes, why did the writer of the New Testament Epistle to the Colossians say, "Let no man judge you in respect of the new moon"? A competent scholar, in recognising this consociation of Hebrew religion with the moon's phases, rightly ascribes to it an earlier origin. Says Ewald: "To connect the annual festivals with the full moon, and to commence them in the evening, as though greeting her with a glad shout, was certainly a primitive custom, both among other races and in the circle of nations from which in the earliest times Israel sprang."[149] And the Bishop of Derry remarks: "To a religious Hebrew it was rather the moon than the sun which marked the seasons, as the calendar of the Church was regulated by it."[150] We have sought to place this Hebrew luniolatry beyond dispute, because so many Christians have supposed that "the chosen people" lived in unclouded light, and "the uncovenanted heathen" in outer and utter darkness.

Passing on we find that "in Pontus and Phrygia were temples to *Meen*, and Homer says *Meen* presides over the months, whilst in the Sanskrit *Mina*, we see her connected with the Fish and Virgin. It is not improbable that the great Akaimenian race, as worshipping and upholding sun and moon faiths, were called after *Meni*, the moon."[151] Among the Arabians the moon was the great divinity, as may be learned from Pocock's *Specimen Historiae Arabum*; Prideaux's *Connection*; Gibbon's *Decline and Fall of the Roman Empire*; and Sale's *Preliminary Discourse* to his translation of the *Koran*. Tiele says: "The ancient religion of the Arabs rises little higher than animistic polydaemonism. The names Itah and Shamsh, the sun god, occur among all the Semitic peoples; Allât, or Alilât, and Al-Uzza, as well as the triad of moon goddesses to which these last belong, are common to several, and the deities which bear them are reckoned among the chief."[152] The Saracens called the moon *Cabar*, the great; and its crescent is the religious symbol of the Turks to this day. Tradition says that "Philip, the father of Alexander, meeting with great difficulties in the siege of Byzantium, set the workmen to undermine the walls, but a crescent moon discovered the design, which miscarried; consequently the Byzantines erected a statue to Diana, and the crescent became the symbol of the state." Dr. Brewer, who cites this story, adds: "Another legend is that Othman, the sultan, saw in a vision a crescent moon, which kept increasing till its horns extended from east to west, and he adopted the crescent of his dream for his standard, adding the motto, *Donec repleat orbem*."[153] Schlegel mentions the story that Mahomet "wished to pass with his disciples as a person transfigured in a supernatural light, and that the credulity of his followers saw the moon, or the moon's light, descend upon him, pierce his garments, and replenish him. That veneration for the moon which still forms a national or rather religious characteristic of the Mahometans, may perhaps have its foundation in the elder superstition, or pagan idolatry of the Arabs."[154] No doubt this last sentence contains the true elucidation of the crescent. For astrolatry lives in the east still. The *Koran* may expressly forbid the practice, saying: "Bend not in adoration to the sun or moon";[155] yet, "monotheist as he is, the Moslem still claps his hands at sight of the new moon, and says a prayer."[156]

We come next to the Persians, whom Herodotus accuses of adoring the sun and moon. But, as Gibbon says, "the Persians of every age have denied the charge, and explained the equivocal conduct, which might appear to give colour to it."[157] It will certainly require considerable explanation to free from lunar idolatry the following passage, which we find in the *Zend Avesta*: "We sacrifice unto the new moon, the holy and master of holiness: we sacrifice unto the full moon, the holy and master of holiness."[158] Unquestionably the Persian recognised the Lord of Light *in* the ordinances of heaven; and therefore his was superior to many forms of blind idol-worship. So far we may accept Hegel's interpretation of the *Zend* doctrine. "Light is the *body of Ormuzd*; thence the worship of fire, because Ormuzd is present in all light; but he is not the sun or moon itself In these the Persians venerate only the light, which is Ormuzd."[159] In fact, we owe to the Persians a valuable testimony to the God in whom is no darkness at all. "The prayer of Ajax was for light"; and we too little feel the Fire which burns and shines beyond the stars.

In Central India the sun and moon are worshipped by many tribes, as the Khonds, Korkús, Tunguses, and Buraets. The Korkús adore the powers of nature, as the gods of the tiger, bison, the hill, the cholera, etc., "but these are all secondary to the sun and the moon, which among this branch of the Kolarian stock, as among the Kols in the far east, are the principal objects of adoration."[160a]

"Although the Tongusy in general worship the sun and moon, there are many exceptions to this observation. I have found intelligent people among them, who believed that there was a being superior to both sun and moon; and who created them and all the world."[160b] This last sentence we read with gratitude, but not with surprise. There is some good in all, if there seem to be all good in some.

"The aboriginal tribes in the Dekkan of India also acknowledge the presence of the sun and moon by an act of reverence."[161]

The inhabitants of the island of Celebes, in the East Indian Archipelago, "formerly acknowledged no gods but the sun and the moon, which were held to be eternal. Ambition for superiority made them fall out."[162] According to Milton, ambition created unpleasantness in the Hebrew heaven.

In Northern Asia the moon had adoring admirers among the Samoyedes, the Morduans, the Tschuwasches, and other tribes. This is stated by Sir John Lubbock.[163] Lord Kames says: "The people of Borneo worship the sun and moon as real divinities. The Samoides worship both, bowing to them morning and evening in the Persian manner."[164] The *Samoides* are the "salmon-eaters" of Asia.

Moon-worship in China is of ancient origin, and exists in our own time. Professor Legge tells us that the primitive *shih* "is the symbol for manifestation and revelation. The upper part of it is the same as that in the older form of Tî, indicating 'what is above'; but of the three lines below I have not found a satisfactory account. Hsü Shăn says they represent 'the sun, moon, and stars,' and that the whole symbolizes 'the indications by these bodies of the will of Heaven! Shih therefore tells us that the Chinese fathers believed that there was communication between heaven and men. The idea of revelation did not shock them. The special interpretation of the strokes below, however, if it were established, would lead us to think that even then, so far back, there was the commencement of astrological superstition, and also, perhaps, of Sabian worship."[165] Sabianism, as most readers are aware, is the adoration of the armies of heaven: the word being derived from the Hebrew *tzaba*, a host. Dr. Legge leaves Chinese Sabianism in some doubt, in the above quotation; but later on he speaks of the spirits associated with the solstitial worship, whose intercession was thus secured, "I, the emperor of the Great Illustrious dynasty, have respectfully prepared this paper, to inform the spirit of the sun, the spirit of the moon, the spirits of the five planets, of the constellations of the zodiac, and of all the stars in all the sky," and so on: and the professor adds: "This paper shows how there had grown up around the primitive monotheism of China the recognition and worship of a multitude of celestial and terrestrial spirits."[166] This is ample evidence to prove moon-worship. True, these celestial beings were "but ministering spirits," and the "monotheism remained." There was no *henotheism*, no worship of several *single* supreme deities: *One only* was supreme. So among the Hebrews, Persians, Hindoos, there was one only God; and yet they offered prayers and sacrifices to heaven's visible and innumerable host. When we come to modern China we shall find some very remarkable celebrations taking place, which throw sunlight upon these ancient mists. Meanwhile to strengthen our position, we may draw additional support from each of the three great stages reached in the progress of Chinese religion: namely, Confucianism, Taoism, and Buddhism. Dr. Edkins describes them as the moral, materialistic, and metaphysical systems, standing at the three corners of a great triangle.[167] The god of Confucianism is *Shang-tî* or *Shang-te*. And with the universal anthropomorphism "Shang-te is the great father of gods and men: Shang-te is a gigantic man."[168] Again "Heaven is a great man, and man is a little heaven."[169] And now what does Confucianism say of moon-worship? "The sun and moon being the chief objects of veneration to the most ancient ancestors of the Chinese, they translated the soul of their great father heaven or the first man (Shang-te) to the sun, and the soul of their great mother earth or the first woman (the female half of the first man) to the moon."[170] In Taoism there is no room for question. Dr. Legge says that it had its Chang and Liû, and "many more gods, supreme gods, celestial gods, great gods, and divine rulers."[171] And Dr. Edkins writes: "The Taouist mythology resembles, in several points, that of many heathen nations. Some of its divinities personate those beings that are supposed to reside in the various departments of nature. Many of the stars are worshipped as gods."[172] Buddhism not only supplies further evidence, it also furnishes a noteworthy instance of mythic transformation. Sakchi or Sasi, the moon, is literally one who made a sacrifice. This refers to the legend of the hare who gave himself to feed the god. The wife of Indra adopted the hare's name, and was herself called Sasi. "The Tantra school gave every deity its Sakti or consort, and speculation enlarged the meaning of the term still further, making it designate female energy or the female principle."[173] Buddhism, then, the popular religion in China at the present day, the religion which Dr. Farrar ventures to call "atheism fast merging into idolatry,"[174] is not free from the nature worship which deifies the moon. But Buddhism, like most other imperfect systems, has precious gold mixed with its dross; and at the expense of a digression we delight to quote the statement of a recent writer, who says: "There is no record, known to me, in the whole of the long history of Buddhism, throughout the many countries where its followers have been for such lengthened periods supreme, of any persecution by the Buddhists of the followers of any other faith."[175] How glad we should feel if we could assert the same of the Christian Church!

We come at once to those celebrations which still take place in China, and illustrate the worship of the moon. The festival of *Yuĕ-Ping*—which is held annually during the eighth month, from the first day when the moon is new, to the fifteenth, when it is full—is of high antiquity and of deep interest. Dr. Morrison says that "the custom of civil and military officers going on the first and fifteenth of every moon to the civil and military temples to burn incense, began in the time of the Lŭh Chaon," which would be not far from A.D. 550. Also that the "eighth month, fifteenth day, is called Chung-tsew-tsëë. It is said that the Emperor Ming-hwang, of the dynasty Tang, was one night led to the palace of the moon, where he saw a large assembly of Chang-go-sëën-neu—female divinities playing on instruments of music. Persons now, from the first to the fifteenth, make cakes like the moon, of various sizes, and paint figures upon them: these are called Yuĕ-ping, 'mooncakes.' Friends and relations pay visits, purchase and present the cakes to each other, and give entertainments. At full moon they spread out oblations and make prostrations to the moon."[176] Dennys writes: "The fifteenth day of the eighth month is a day on which a ceremony is performed by the Chinese, which of all others we should least expect to find imitated among ourselves. Most people resident in China have seen the moon-cakes which so delight the heart of the Chinese during the eighth month of every year. These are made for an autumnal festival often described as 'congratulating' or 'rewarding' the moon. The moon, it is well known, represents the female principle in Chinese celestial cosmogony, and she is further supposed to be inhabited by a multitude of beautiful females; the cakes made in her honour are therefore veritable offerings to the Queen of the Heavens. Now in a part of Lancashire, on the banks of the Ribble, there exists a precisely similar custom of making cakes in honour of the 'Queen of Heaven,'—a relic, in all probability, of the old heathen worship which was the common fount of the two customs."[177] Witness is also borne to this ceremony by a well-known traveller. "We arrived at Chaborté on the fifteenth day of the eighth moon, the anniversary of great rejoicings among the Chinese. This festival, known as the *Yuĕ-Ping* (loaves of the moon), dates from the remotest antiquity. Its original purpose was to honour the moon with superstitious rites. On this solemn day, all labour is suspended; the workmen receive from their employers a present of money, every person puts on his best clothes; and there is merry-making in every family. Relations and friends interchange cakes of various sizes, on which is stamped the image of the moon; that is to say, a hare crouching amid a small group of trees."[178] And Doolittle says: "It is always full moon on the fifteenth of every Chinese month; and, therefore, for several days previous, the evenings are bright, unless it happens to be cloudy, which is not often the case. The moon is a prominent object of attention and congratulation at this time. At Canton, it is said, offerings are made to the moon on the fifteenth. On the following day, young people amuse themselves by playing what is called '*pursuing*,' or '*congratulating*' the moon. At this city[Fuhchau], in the observance of this festival, the expression '*rewarding the moon*' is more frequently used than 'congratulating the moon.' It is a common saying that there is 'a white rabbit in the moon pounding out rice.' The dark and the white spots on the moon's face suggest the idea of that animal engaged in the useful employment of shelling rice. The notion is prevalent that the moon is inhabited by a multitude of beautiful females, who are called by the name of an ancient beauty who once visited that planet; but how they live, and what they do, is not a matter of knowledge or of common fame. To the question, 'Is the moon inhabited?' discussed by some Western philosophers, the Chinese would answer in the affirmative. Several species of trees and flowers are supposed to flourish in the moon. Some say that, one night in ancient times, one of the three souls of the originator of theatrical plays rambled away to the moon and paid a visit to the Lunar Palace. He found it filled with Lunarians engaged in theatrical performances. He is said to have remembered the manner of conducting fashionable theatres in the moon, and to have imitated them after his return to this earth. About the time of the festival of the middle of autumn, the bake shops provide an immense amount and variety of cakes: many of them are circular, in imitation of the shape of the moon at that time, and are from six to twelve inches in diameter. Some are in the form of a pagoda, or of a horse and rider, or of a fish, or other animals which please and cause the cake to be readily sold. Some of these 'moon-cakes' have a white rabbit, engaged with his pounder, painted on one side, together with a lunar beauty, and some trees or shrubs; on others are painted gods or goddesses, animals, flowers, or persons, according to fancy."[179]

If we turn now to Jeremiah vii. 18, and read there, "The women knead dough, to make cakes to the Queen of Heaven, and to pour out drink offerings unto other gods," and remember that, according to Rashi, these cakes of the Hebrews had the image of the god or goddess stamped upon them, we are in view of a fact of much interest. We are so unaccustomed to think that our peasants in Lancashire can have anything in common with the Chinese five thousand miles away, and with the Jews of two thousand five hundred years ago, that to many these moon-cakes will give a genuine surprise. But this is not all. Other analogies appear between Buddhist and Christian rites, such as those mentioned by Dr. Medhurst. "The very titles of their intercessors, such as 'goddess of mercy,' 'holy mother,' 'queen of heaven,' with the image of a virgin, having a child in her arms, holding a cross, are all such striking coincidences, that the Catholic missionaries were greatly stumbled at the resemblance between the Chinese worship and their own, when they came over to convert the natives to Christianity."[180] It is for the philosophical historian to show, if possible, whether these Chinese ceremonies are copies of Christian or Hebrew originals; or whether, many of our own Western forms with others of Oriental character, are not transcripts of primitive faiths now well-nigh forgotten in both East and West. The hot cross buns of Good Friday, at first sight, have little relevancy to moon worship, and those who eat them suppose they were originated to commemorate the Christian Sacrifice; but we know that the cross was a sacred symbol with the earliest Egyptians, for it is carved upon their imperishable records; we know too that *bun* itself is ancient Greek, and that Winckelmann relates the discovery at Herculaneum of two perfect buns, each marked with a cross: while the *boun* described by Hesychius was a cake with a representation of *two horns*. Incredible as it may seem to some, the cross bun in its origin had nothing to do with an event with which it is in England identified; it probably commemorates the worship of the moon. In passing from China, we may also note the influence of that sexuality of which we have spoken before. Dr. Medhurst remarks: "The principle of the Chinese cosmogony seems to be founded on a sexual system of the universe."[181]

Dr. Prichard tells us that among the Japanese "sacred festivals are held at certain seasons of the year and at changes of the moon." Also, "It appears that *Sin-too*, or original Japanese religion, is merely a form of the worship of material objects, common to all the nations of Northern Asia, which, among the more civilized tribes, assumes the aspect of mythology."[182]

From Asia we come to Africa, and to Egypt, that wonderful land with a lithographed history at least five thousand years old; a land that basked in the sunshine of civilization and culture when nearly the whole world without was in shadow and gloom. The mighty pyramid of Gizeh still stands, a monument of former national greatness, and a marvel to the admirer of sublimity in design and perfection in execution. "The setting of the sides to the cardinal points is so exact as to prove that the Egyptians were excellent observers of the elementary facts of astronomy."[183] But they went farther. Diodorus says: "The first generation of men in Egypt, contemplating the beauty of the superior world, and admiring with astonishment the frame and order of the universe, judged that there were two chief gods that were eternal, that is to say, the sun and the moon, the first of which they called *Osiris*, and the other *Isis*."[184] This passage is proof that the Greeks and Romans had a very limited acquaintance with Egyptian mythology; for the historian was indubitably in error in supposing Osiris and Isis to be sun and moon. But he was right in calling the sun and moon the first gods of the Egyptians. Rawlinson says: "The Egyptians had two moon-gods, Khons or Khonsu, and Tet or Thoth."[185] Dr. Birch has translated an inscription relating to Thoth, which reads: "All eyes are open on thee, and all men worship thee as a god."[186] And M. Renouf says: "The Egyptian god Tehuti is known to the readers of Plato under the name of Thōyth. He represents the moon, which he wears upon his head, either as crescent or as full disk."[187] The same learned Egyptologist tells us that Khonsu or Chonsu was one of the triad of Theban gods, and was the moon one of his attributes being the reckoner of time.[188] Of the former divinity, Rawlinson relates an instructive myth. "According to one legend Thoth once wrote a wonderful book, full of wisdom and science, containing in it everything relating to the fowls of the air, the fishes of the sea, and the four-footed beasts of the mountains. The man who knew a single page of the work could charm the heaven, the earth, the great abyss, the mountains and the seas. This marvellous composition he inclosed in a box of gold, which he placed within a box of silver; the box of silver within a box of ivory and ebony, and that again within a box of bronze; the box of bronze within a box of brass; and the box of brass within a box of iron; and the book, thus guarded, he threw into the Nile at Coptos. The fact became known, and the book was searched for and found. It gave its possessor vast knowledge and magical power, but it always brought on him misfortune. What became of it ultimately does not appear in the manuscript from which this account is taken; but the moral of the story seems to be the common one, that unlawful knowledge is punished by all kinds of calamity."[189] There is also a story of the moon-god Chonsu, which is worthy of repetition. Its original is in the *Bibliothèque Nationale* at Paris, and for its first translation we are indebted to Dr. Birch, of the British Museum.[190] A certain Asiatic princess of Bechten, wherever that was, was possessed by a spirit. Being connected, through her sister's marriage, with the court of Egypt, on her falling ill, an Egyptian practitioner was summoned to her aid. He declared that she had a demon, with which he himself was unable to cope. Thereupon the image of the moon-god Chonsu was despatched in his mystic ark, for the purpose of exorcising the spirit and delivering the princess. The demon at once yielded to the divine influence; and the king of Bechten was so delighted that he kept the image in his possession for upwards of three years. In consequence of an alarming dream he then sent him back to Egypt with presents of great value. Whatever evil powers the moon may have exerted since, we must credit him with having once ejected an evil spirit and prolonged a royal life.

Returning to Thoth, we find the following valuable hints in the great work of Baron Bunsen:—"The connection between Tet and the moon may allude, according to Wilkinson, to the primitive use of a lunar year. The ancients had already remarked that the moon in Egyptian was masculine, not feminine, as the Greeks and Romans generally made it. Still we have no right to suppose a particular moon-god, separate from Thoth. We meet with a deity called after the moon (Aah) either as a mere personification, or as Thoth, in whom the agency of the moon and nature become a living principle. We find him so represented in the tombs of the Ramesseum, opposite to Phre; a similar representation in Dendyra is probably symbolical. According to Champollion he is often seen in the train of Ammon, and then he is Thoth. He makes him green, with the four sceptres and cup of Ptah, by the side of which, however, is a sort of Horus curl, the infantine lock, as child or son. In the inscriptions there is usually only the crescent, but on one occasion the sign *nuter* (god) is added. In the tombs a moon-god is represented sitting on a bark, and holding the sceptre of benign power, to whom two Cynocephali are doing homage, followed by the Crescent and Nuter god. Lastly, the same god is found in a standing posture, worshipped by two souls and two Cynocephali."[191]

With these "dog-headed" worshippers of the moon may be associated another animal that from an early date has been connected with the luminaries of the day and night. We saw that the Australian moon-myth of Mityan was of a native cat. Renouf says: "It is not improbable that the cat, in Egyptian *mäu*, became the symbol of the Sun-god, or Day, because the word mäu also means light."[192] Charles James Fox, with no thought of Egyptian, told the Prince of Wales that "cats always prefer the sunshine." The native land of this domestic pet, or nuisance, is certainly Persia, and some etymologists assign *pers* as the origin of *puss*. Be this as it may, the pupil of a cat's eye is singularly changeable, dilating from the narrow line in the day-time to the luminous orb in the dark. On this account the cat is likened to the moon. But in Egypt feline eyes shine with supernatural lustre. Mr. Hyde Clarke tells us that "the mummies of cats, which Herodotus saw at Bubastis, attested then, as they do now, to the dedication of the cat to Pasht, the moon, and the veneration of the Egyptians for this animal. The cat must have been known to man, and have been named at least as early as the origin of language. The superstition of its connection with the moon is also of pre-historic date, and not invented by the Egyptians. According to Plutarch, a cat placed in a lustrum denoted the moon, illustrating the mutual symbology. He supposes that this is because the pupils of a cat's eyes dilate and decrease with the moon. The reason most probably depends, as before intimated, on another phenomenon of periodicity corresponding to the month. Dr. Rae has, however, called my attention to another possible cause of the association, which is the fact that the cat's eyes glisten at night or in the dark. It is to be observed that the name of the sun in the Malayan and North American languages is the day-eye, or sky-eye, and that of the moon the night-eye."[193] Our own daisy, too, is the *day's eye*, resembling the sun, and opening its little pearly lashes when the spring wakes to newness of life.

The Nubians "pay adoration to the moon; and that their worship is performed with pleasure and satisfaction, is obvious every night that she shines. Coming out from the darkness of their huts, they say a few words upon seeing her brightness, and testify great joy, by motions of their feet and hands, at the first appearance of the new moon."[194] The Shangalla worship the moon, and think that "a star passing near the horns of the moon denotes the coming of an enemy."[195] In Western Africa moon-worship is very prevalent. Merolla says: "They that keep idols in their houses, every first day of the moon are obliged to anoint them with a sort of red wood powdered. At the appearance of every new moon, these people fall on their knees, or else cry out, standing and clapping their hands, 'So may I renew my life as thou art renewed.'"[196]

H. H. Johnston, Esq., F.Z.S., F.R.G.S., who had just returned from the region of the Congo, related the following curious incident before the Anthropological Institute, in January, 1884. It looks remarkably like a relic of ancient worship, which gave the fruit of the body for the sin of the soul, and committed murder on earth to awaken mercy in heaven! "At certain villages between Manyanga and Isangila there are curious eunuch dances to celebrate the new moon, in which a white cock is thrown up into the air alive, with clipped wings, and as it falls towards the ground it is caught and plucked by the eunuchs. I was told that originally this used to be a human sacrifice, and that a young boy or girl was thrown up into the air and torn to pieces by the eunuchs as he or she fell, but that of late years slaves had got scarce or manners milder, and a white cock was now substituted."[197]

The Mandingoes are more attracted to the varying moon than to the sun. "On the first appearance of the new moon, which they look upon to be newly created, the Pagan natives, as well as Mahomedans, say a short prayer; and this seems to be the only visible adoration which the Kaffirs offer up to the Supreme Being." The purport of this prayer is "to return thanks to God for His kindness through the existence of the past moon, and to solicit a continuation of His favour during that of the new one."[198] Park writes on another page: "When the fast month was almost at an end, the Bushreens assembled at the Misura to watch for the appearance of the new moon; but the evening being rather cloudy, they were for some time disappointed, and a number of them had gone home with a resolution to fast another day, when on a sudden this delightful object showed her sharp horns from behind a cloud, and was welcomed with the clapping of hands, beating of drums, firing muskets, and other marks of rejoicing."[199] The Makololo and Bechuana custom of greeting the new moon is curious. "They watch most eagerly for the first glimpse of the new moon, and when they perceive the faint outline after the sun has set deep in the west, they utter a loud shout of 'Kuā!' and vociferate prayers to it."[200] The degraded Hottentots have not much improved since Bory de St. Vincent described them as "brutish, lazy, and stupid," and their worship of the moon is still demonstrative, as when Kolben wrote: "These dances and noises are religious honours and invocations to the moon. They call her *Gounja*. The Supreme they call *Gounja-Gounja*, or *Gounja Ticquoa*, the god of gods, and place him far above the moon. The moon, with them, is an inferior visible god—the subject and representation of the High and Invisible. They judge the moon to have the disposal of the weather, and invoke her for such as they want. They assemble for the celebration of her worship at full and change constantly. No inclemency of the weather prevents them. And their behaviour at those times is indeed very astonishing. They throw their bodies into a thousand different distortions, and make mouths and faces strangely ridiculous and horrid. Now they throw themselves flat on the ground, screaming out a strange, unintelligible jargon. Then jumping up on a sudden, and stamping like mad (insomuch that they make the ground shake), they direct, with open throats, the following expressions, among others, to the moon: '*I salute you; you are welcome. Grant us fodder for our cattle and milk in abundance.*' These and other addresses to the moon they repeat over and over, accompanying them with dancing and clapping of hands. At the end of the dance they sing '*Ho! Ho! Ho! Ho!*' many times over, with a variation of notes; which being accompanied with clapping of hands makes a very odd and a very merry entertainment to a stranger."[201] In reality they hold a primitive watch-night service; their welcome of the new moon being very similar to our popular welcome of the new year. Nor should it be omitted that the ancient Ethiopians worshipped the moon; and that those who lived above Meroë admitted the existence of eternal and incorruptible gods, among which the moon ranked as a chief divinity.

Descending the Nile and crossing the Mediterranean, we come to Greece.

"The isles of Greece, the isles of Greece
  Where burning Sappho loved and sung,
Where grew the arts of war and peace,
  Where Delos rose, and Phoebus sprung
Eternal summer gilds them yet,
But all, except their sun, is set."[202]

Yes, Pericles and Plato, Sophocles and Pheidias, are dust; and much of their nation's pristine glory has "melted into the infinite azure of the past": but the sun shines as youthful yet as on that eventful day when unwearied he sank in ocean, "loth, and ere his time:

> "So the sun sank, and all the host had rest
> From onset and the changeful chance of war."[203]

Where Phoebus sprang, sprang Phoebe also—the bright and beautiful moon. To a people addicted to the idolatry of perfect form and comeliness, no object could be more attractive than the queen of the night. When Socrates was accused of innovating upon the Greek religion, and of ridiculing the Athenian deities, he replied on his trial, "You strange man, Melêtus, are you seriously affirming that I do not think Helios and Selene to be gods, as the rest of mankind think?"[204] Pausanias, the historian, tells us that in Phocis there was a chapel consecrated to Isis, which of all the places erected by the Greeks to this Egyptian goddess was by far the most holy. It was not lawful for any one to approach this sacred edifice but those whom the goddess had invited by appearing to them for that purpose in a dream.[205] By Isis, as we saw from Diodorus, the Greeks understood the moon. Diana was also one of the Grecian moon-goddesses, but Sir George C. Lewis thinks that this was not till a comparatively late period. The religion of Greece was so mixed up, or made up, with mythology, that for an interpretation of their theogony we must resort to poetry and impersonation. Here again we see the working of sexual anthropomorphism. *Ouranos* espoused *Ge*, and their offspring was *Kronos*; which is but an ancient mode of saying that chronology is the measurement on earth of heavenly motion. Solar and lunar worship was but the recognition in the primitive consciousness of the superior *worth-ship* of these celestial bodies. As Grote says: "To us these now appear puerile, though pleasing fancies, but to our Homeric Greek they seemed perfectly natural and plausible. In his view, the description of the sun, as given in a modern astronomical treatise, would have appeared not merely absurd, but repulsive and impious."[206] What an amount of misunderstanding would be obviated if readers of the Bible would bear this in mind when they meet with erroneous conceptions in Hebrew cosmogony. Grote further says on the same page of his magnificent history: "Personifying fiction was blended by the Homeric Greeks with their conception of the physical phenomena before them, not simply in the way of poetical ornament, but as a genuine portion of their everyday belief." We cannot better conclude our brief glance at ancient Greece than by quoting that splendid comparison from the bard of Chios, which Pope thought "the most beautiful night-piece that can be found in poetry." Pope's own version is fine, but, as a translation, Lord Derby's must be preferred:

> "As when in heaven, around the glittering moon
> The stars shine bright amid the breathless air;
> And every crag and every jutting peak
> Stands boldly forth, and every forest glade
> Even to the gates of heaven is opened wide
> The boundless sky; shines each particular star
> Distinct; joy fills the gazing shepherd's heart."[207]

The Romans had many gods, superior and inferior. The former were the *celestial* deities, twelve in number, among whom was Diana; and the *Dii Selecti*, numbering eight. Of these, one was Luna, the moon, daughter of Hyperion and sister of the Sun.[208] Livy speaks of "a temple of Luna, which is on the Aventine"; and Tacitus mentions, in his Annals, a temple consecrated to the moon. In Horace, Luna is "*siderum regina*";[209] and in Apuleius, "*Regina coeli*,"[210] Bishop Warburton, in his synopsis of Apuleius, speaks of the hopeless condition of *Lucius*, which obliged him to fly to heaven for relief. "The *moon* is in full splendour; and the awful silence of the night inspires him with sentiments of religion." He then purifies himself, and so makes his prayer to the moon, invoking her by her several names, as the celestial *Venus* and *Diana*.[211] This whole section of the *Divine Legation* is worthy of close study.

"The ancient Goths," says Rudbeck ("Atalantis," ii. 609), "paid such regard to the moon, that some have thought that they worshipped her more than the sun."[212]

And of the ancient Germans Grimm says: "That to our remote ancestry the heavenly bodies, especially the sun and moon, were divine beings, will not admit of any doubt."[213] Gibbon, Friedrich Schlegel, and others, say the same.

The Finns worshipped "Kun, the male god of the moon, who corresponded exactly with the Aku, Enizuna, or Itu of the Accadians."[214]

In ancient Britain the moon occupied a high position in the religion of the Druids, who had superstitious rites at the lunar changes, and who are "always represented as having the crescent in their hands."[215] "From the *Penitential* of Theodore, Archbishop of Canterbury, in the seventh century, and the *Confessional* of Ecgbert, Archbishop of York, in the early part of the eighth century, we may infer that homage was then offered to the sun and moon."[216] Again, "There are many proofs, direct and circumstantial, that place it beyond all doubt that the moon was one of the objects of heathen worship in Britain. But under what name the moon was invoked is not discoverable, unless it may have been Andraste, the goddess to whom the British queen Boadicea, with hands outstretched to heaven, appealed when about to engage in battle with the Romans."[217] A writer of the seventeenth century, says: "In Yorkeshire, etc., northwards, some country woemen do-e worship the New Moon on their bare knees, kneeling upon an earthfast stone. And the people of Athol, in the High-lands in Scotland, doe worship the New Moon."[218] Camden writes of the Irish: "Whether or no they worship the moon, I know not; but, when they first see her after the change, they commonly bow the knee, and say the Lord's Prayer; and near the wane, address themselves to her with a loud voice, after this manner: 'Leave us as well as thou foundest us.'"[219] Sylvester O'Halloran, the Irish general and historian, speaking of "the correspondent customs of the Phoenicians and the Irish," adds: "Their deities were the same. They both adored Bel, or the sun, the Moon, and the stars. The house of Rimmon (2 Kings v. 18), which the Phoenicians worshipped in, like our temples of Fleachta, in Meath, was sacred to the moon. The word 'Rimmon' has by no means been understood by the different commentators; and yet by recurring to the Irish (a branch of the Phoenicians) it becomes very intelligible; for *Re* is Irish for the moon, and *Muadh* signifies an image; and the compound word *Reamham* signifies prognosticating by the appearances of the moon. It appears by the life of our great St. Columba, that the Druid temples were *here* decorated with figures of the sun, the moon, and the stars. The Phoenicians, under the name of Bel-Samen, adored the Supreme; and it is pretty remarkable that *to this very day*, to wish a friend every happiness this life can afford, we say in Irish, '*the blessings of Samen and Eel be with you*!' that is, of all the seasons; Bel signifying the sun, and Samhain the moon."[220] And again: "Next to the sun was the moon, which the Irish undoubtedly adored. Some remains of this worship may be traced, even at this day; as particularly borrowing, if they should not have it about them, a piece of silver on the first night of a new moon, as an omen of plenty during the month; and at the same time saying in Irish, 'As you have found us in peace and prosperity, so leave us in grace and mercy.'"[221] Tuathal, the prince to whom the estates (*circa* A.D. 106) swore solemnly "by the sun, moon, and stars," to bear true allegiance, "in that portion of the imperial domain taken from Munster, erected a magnificent temple called Flachta, sacred to the fire of Samhain, and to the Samnothei, or priests of the moon. Here, on every eve of November, were the fires of Samhain lighted up, with great pomp and ceremony, the monarch, the Druids, and the chiefs of the kingdom attending; and from this holy fire, and no other, was every fire in the land first lit for the winter. It was deemed an act of the highest impiety to kindle the winter fires from any other; and for this favour the head of every house paid a Scrubal, or threepence, tax, to the Arch-Druid of Samhain."[222] Another writer mentions another Irish moon-god. "The next heathen divinity which I would bring under notice is St. Luan, *alias* Molua, *alias*

Euan, *alias* Lugidus, *alias* Lugad, and Moling, etc. The foundations, with which this saint under some of his *aliases* is connected, extend over eight counties in the provinces of Ulster, Leinster, and Munster. Luan is to this; day the common Irish word for the moon. We read that there were fifteen saints of the name of Lugadius; and as Lugidus was one of Luan's *aliases,* I have set them all down as representing the moon in the several places where that planet was worshipped as the symbol of Female nature."[223] We have already seen that the moon was the embodiment of the female principle in China, and now we see that the primitive Kelts associated sexuality with astronomy and religion. It but further proves that "one touch of nature makes the whole world kin."

Moreover, to show that former moon-worship still colours our religion, it is not to be overlooked that, as our Christmas festivities are but a continuation of the Roman saturnalia, with their interchanges of visits and presents, so "the Church, celebrating in August the festival of the harvest moon, celebrates at the same time the feast of the Assumption and of the Sacred Heart of the Virgin. And Catholic painters, following the description in the Apocalypse, fondly depict her as 'clothed with the sun, and having the moon under her feet,' and both as overriding the dragon. Even the triumph of Easter is not celebrated until, by attaining its full, the moon accords its aid and sanction. Is it not interesting thus to discover the true note of Catholicism in the most ancient paganisms, and to find that the moon, which for us is incarnate in the blessed Virgin Mary, was for the Syrians and Greeks respectively personified in the virgin Ashtoreth, the queen of heaven, and Diana, or Phoebe, the feminine of Phoebus?"[224]

A recent contributor to one of our valuable serials writes: "I take the following extract from a little book published under the auspices of Dr. Barnardo. It is the 'truthful narrative' of a little sweep-girl picked up in the streets of some place near Brighton, and 'admitted into Dr. Barnardo's Village Home.' 'She had apparently no knowledge of God or sense of His presence. The only thing she had any reverence for was the moon. On one occasion, when the children were going to evening service, and a beautiful moon was shining, one of them pointed to it, exclaiming, 'Oh, mother! look, what a beautiful moon!' Little Mary caught hold of her hand, and cried, 'Yer mustn't point at the blessed moon like that; and yer mustn't talk about it!' Was it from constantly sleeping under hedges and in barns, and waking up and seeing that bright calm eye looking at her, that some sense of a mysterious Presence had come upon the child?"[225] To this query, the answer we think should be negative. The cause more likely was that she had heard the common tradition which is yet current in East Lancashire, Cumberland, and elsewhere, that it is a sin to point at the moon. Certain old gentlemen, who ought to be better informed, still touch their hats, and devout young girls in the country districts still curtsey, to the new moon, as an act of worship.

The American races practise luniolatry very generally. The Dakotahs worship both sun and moon. The Delaware and Iroquois Indians sacrifice to these orbs, and it is most singular that "they sacrifice to a hare, because, according to report, the first ancestor of the Indian tribes had that name." But, although they receive in a dream as their tutelar spirits, the sun, moon, owl, buffalo, and so forth, "they positively deny that they pay any adoration to these subordinate good spirits, and affirm that they only worship the true God, through them."[226] This reminds us of some excellent remarks made by one whose intimate acquaintance with North American Indians entitled him to speak with authority. We have seen from Dr. Legge's writings that though the Chinese worshipped a multitude of celestial spirits, "yet the monotheism remained." Mr. Catlin will now assure us that though the American Indians adore the heavenly bodies, they recognise the Great Spirit who inhabits them all. These are his words: "I have heard it said by some very good men, and some who have even been preaching the Christian religion amongst them, that they have no religion—that all their zeal in their worship of the Great Spirit was but the foolish excess of ignorant superstition—that their humble devotions and supplications to the sun and the moon, where many of them suppose that the Great Spirit resides, were but the absurd rantings of idolatry. To such opinions as these I never yet gave answer, nor drew other instant inferences from them, than that, from the bottom of my heart, I pitied the persons who gave them."[227] Mr. Catlin undoubtedly was right, as the Apostle Paul was right, when he acknowledged that the Athenians worshipped the true God, albeit in ignorance. At the same time, though idolatry is in numberless instances nothing more than the use of media and mediators, in seeking the One, Invisible, Absolute Spirit, it is so naturally abused by sensuous beings who rest in the concrete, that no image worshipper is free from the propensity to worship the creature more than the Creator, and to forget the Essence in familiarity with the form. The perfection of worship, we conceive, is pure theism; but how few are capable of breathing in such a supersensuous air! Men must have their "means of grace," their visible symbols, their holy waters and consecrated wafers, their crucifixes and talismans, their silver shrines and golden calves. "These be thy gods, O Israel."

"The Ahts undoubtedly worship the sun and the moon, particularly the full moon, and the sun while ascending to the zenith. Like the Teutons, they regard the moon as the husband, and the sun as the wife; hence their prayers are more generally addressed to the moon, as being the superior deity. The moon is the highest of all the objects of their worship; and they describe the moon—I quote the words of my Indian informant—as looking down upon the earth in answer to prayer, and as seeing everybody."[228] Of the Indians of Vancouver Island, another writer says: "The moon is among all the heavenly bodies the highest object of veneration. When working at the settlement at Alberni in gangs by moonlight, individuals have been observed to look up to the moon, blow a breath, and utter quickly the word, '*Teech! teech!*' (health, or life). Life! life! this is the great prayer of these people's hearts."[229] "Among the Comanches of Texas, the sun, moon, and earth are the principal objects of worship." The Kaniagmioutes consider the moon and sun to be brother and sister.[230]

Meztli was the moon as deified by the Mexicans. In Teotihuacan, thirty miles north of the city of Mexico, is the site of an ancient city twenty miles in circumference. Near the centre of this spot stand the Pyramid of the Sun and the Pyramid of the Moon. The Pyramid of the Sun has a base 682 feet long and is 180 feet high (the Pyramid of Cheops is 728 feet at the base, and is 448 feet in height). The Pyramid of the Moon is rather less, and is due north of that of the Sun.[231] No doubt the philosophy of all pyramids would show that they embody the uplifting of the human soul towards the Heaven-Father of all.

In Northern Mexico still "the Ceris superstitiously celebrate the new moon."[232] This luniolatry the Abbé Brasseur de Bourbourg explains by a novel theory. He holds that the forefathers of American civilization lived in a certain Crescent land in the Atlantic that a physical catastrophe destroyed their country whereupon the remnant that was saved commemorated their lost land by adopting the moon as their god.[233] "The population of Central America," says the Vicomte de Bussierre, "although they had preserved the vague notion of a superior eternal God and Creator, known by the name Teotl, had an Olympus as numerous as that of the Greeks and the Romans. It would appear that the inhabitants of Anahuac joined to the idea of a supreme being the worship of the sun and the moon, offering them flowers, fruits, and the first fruits of their fields."[234] Dr. Reville bids us "note that the ancient Central-American cultus of the sun and moon, considered as the two supreme deities, was by no means renounced by the Aztecs."[235] Regarding this remarkable race, a writer in the *Quarterly Review* for April, 1883, says: "Even the Chaldaeans were not greater astrologers than the Aztecs, and we need no further proof that the heavenly bodies were closely and accurately observed, than we find in the fact that the true length of the tropical year had been ascertained long before scientific instruments were even thought of. Their religious festivals were regulated by the movements of these bodies; but with their knowledge was mingled so vast a mass of superstition, that it is difficult to discern a gleam of light through the thick darkness." "The Botocudos of Brazil held the moon in high veneration, and attributed to her influence the chief phenomena in nature."[236] The Indian of the Coroados tribe in Brazil, "chained to the present, hardly ever raises his eyes to the starry firmament. Yet he is actuated by a certain awe of some constellations, as of everything that indicates a spiritual connection of things. His chief attention, however, is not directed to the sun, but to the moon; according to which he calculates time, and from which he is used to deduce good and evil."[237]

The celebrated Abipones honour with silver altars and adoration the moon, which they call the consort of the sun, and certain stars, which they term the handmaids of the moon: but their most singular idea is that the Pleiades represent their grandfather; and "as that constellation disappears at certain periods from the sky of South America, upon such occasions they suppose that their grandfather is sick, and are under a yearly apprehension that he is going to die; but as soon as those seven stars are again visible in the month of May, they welcome their grandfather, as if returned and restored from sickness, with joyful shouts, and the festive sound of pipes and trumpets, congratulating him on the recovery of his health."[238]

The Peruvians "acknowledge no other gods than the Pachacamac, who is the supreme, and the Sun, who is inferior to him, and the Moon, who is his sister and wife."[239] In the religion of the Incas the idol (huaco) of the Moon was in charge of women, and when it was brought from the house of the Sun, to be worshipped, it was carried on their shoulders, because they said "it was a woman, and the figure resembled one."[240] *Pachacamac*, the great deity mentioned above, signifies "earth-animator."

Prescott, in describing the temple of the Sun, at Cuzco in Peru, tells us that "adjoining the principal structure were several chapels of smaller dimensions. One of them was consecrated to the Moon, the deity held next in reverence, as the mother of the Incas. Her effigy was delineated in the same manner as that of the Sun, on a vast plate that nearly covered one side of the apartment. But this plate, as well as all the decorations of the building, was of silver, as suited to the pale, silvery light of the beautiful planet."[241]

In the far-off New Hebrides the Eramangans "worship the moon, having images in the form of the new and full moons, made of a kind of stone. They do not pray to these images, but cleave to them as their protecting gods."[242]

We have now circumnavigated the globe, touching at many points, within many degrees of latitude and longitude. But everywhere, among men of different literatures and languages, colours and creeds, we have discovered the worship of the moon. No nation has outgrown the practice, for it obtains among the polished as well as the rude. One thing, indeed, we ought to have had impressed upon our minds with fresh force; namely, that we often draw the lines of demarcation too broad between those whom we are pleased to divide into the civilized and the savage. Israelite and heathen, Grecian and barbarian, Roman and pagan, enlightened and benighted, saintly and sinful, are fine distinctions from the Hebrew, Greek, Roman, enlightened, and saintly sides of the question; but they often reflect small credit upon the wisdom and generosity of their authors. The antipodal Eramangan who cleaves to his moon image for protection may be quite equal, both intellectually and morally, with the Anglo-Saxon who still wears his amulet to ward off disease, or nails up his horse-shoe, as Nelson did to the mast of the *Victory*, as a guarantee of good luck. Sir George Grey has written: "It must be borne in mind, that the native races, who believed in these traditions or superstitions, are in no way deficient in intellect, and in no respect incapable of receiving the truths of Christianity; on the contrary, they readily embrace its doctrines and submit to its rules; in our schools they stand a fair comparison with Europeans; and, when instructed in Christian truths, blush at their own former ignorance and superstitions, and look back with shame and loathing upon their previous state of wickedness and credulity."[243]

The Moon a Water-Deity

We design this chapter to be the completion of moon-worship, and at the same time an anticipation of those lunary superstitions which are but scattered leaves from luniolatry, the parent tree. If the new moon, with its waxing light, may represent the primitive nature-worship which spread over the earth; and the full moon, the deity who is supposed to regulate our reservoirs and supplies of water: the waning moon may fitly typify the grotesque and sickly superstition, which, under the progress of radiant science and spiritual religion, is readier every hour to vanish away.

"The name Astarte was variously identified with the moon, as distinguished from the sun, or with air and water, as opposed in their qualities to fire. The name of this goddess represented to the worshipper the great female parent of all animated things, variously conceived of as the moon, the earth, the watery element, primeval night, the eldest of the destinies."[244] It is worthy of note that Van Helmont, in the seventeenth century, holds similar language. His words are, "The moon is chief over the night darkness, rest, death, and the waters."[245] It is also remarkable that in the language of the Algonquins of North America the ideas of night, death, cold, sleep, water, and moon are expressed by one and the same word.[246] In the oriental mythology "the connection between the moon and water suggests the idea that the moon produces fertility and freshness in the soil."[247] "Al Zamakhshari, the commentator on the Koran, derives *Manah* (one of the three idols worshipped by the Arabs before the time of Mohammad) from the root 'to flow,' because of the blood which flowed at the sacrifices to this idol, or, as Millius explains it, because the ancient idea of the moon was that it was a star full of moisture, with which it filled the sublunary regions."[248] The Persians held that the moon was the cause of an abundant supply of water and of rain, and therefore the names of the most fruitful places in Persia are compounded with the word *mâh*, "moon"; "for in the opinion of the Iranians the growth of plants depends on the influence of the moon."[249] In India "the moon is generally a male, for its most popular names, *Candras*, *Indus*, and *Somas*, are masculine; but as Somas signifies ambrosia, the moon, as giver of ambrosia, soon came to be considered a milk-giving cow; in fact, moon is one among the various meanings given in Sanskrit to the word Gâus (cow). The moon, Somas, who illumines the nocturnal sky, and the pluvial sun, Indras, who during the night, or the winter, prepares the light of morn, or spring, are represented as companions; a young girl, the evening, or autumnal twilight, who goes to draw water towards night, or winter, finds in the well, and takes to Indras, the ambrosial moon, that is, the Somas whom he loves. Here are the very words of the Vedic hymn: 'The young girl, descending towards the water, found the moon in the fountain, and said: I will take you to Indras, I will take you to Çakras; flow, O moon, and envelop Indras.'"[250] Here in India we again find our old friend "the frog in the moon." "It is especially Indus who satisfies the frog's desire for rain. Indus, as the moon, brings or announces the Somas, or the rain; the frog, croaking, announces or brings the rain; and at this point the frog, which we have seen identified at first with the cloud, is also identified with the pluvial moon."[251] This myth is not lacking in involution.

In China "the moon is regarded as chief and director of everything subject in the kosmic system to the Yin[feminine] principle, such as darkness, the earth, female creatures, water, etc. Thus Pao P'ah Tsze declares with reference to the tides: 'The vital essence of the moon governs water: and hence, when the moon is at its brightest, the tides are high.'"[252] According to the Japanese fairy tale the moon was to "rule over the new-born earth and the blue waste of the sea, with its multitudinous salt waters."[253] Thus we see that throughout Asia, "as lord of moisture and humidity, the moon is connected with growth and the nurturing power of the peaceful night."[254]

Of the kindred of the Pharaohs, Plutarch observes: "The sun and moon were described by the Egyptians as sailing round the world in boats, intimating that these bodies owe their power of moving, as well as their support and nourishment, to the principle of humidity" (Plut. de Isid. s. 34): which statement Sir J. Gardner Wilkinson says is confirmed by the sculptures. The moon-god Khons bears in his hands either a palm-branch or "the Nilometer." When the Egyptians sacrificed a pig to the moon, "the first sacred emblem they carried was a *hydria*, or water-pitcher." At another festival the Egyptians "marched in procession towards the sea-side, whither likewise the priests and other proper officers carried the sacred chest, inclosing a small boat or vessel of gold, into which they first poured some fresh water; and then all present cried out with a loud voice 'Osiris is found.' This ceremony being ended, they threw a little fresh mould, together with rich odours and spices, into the water, mixing the whole mass together, and working it up into a little image in the shape of a crescent. The image was afterwards dressed and adorned with a proper habit, and the whole was intended to intimate that they looked upon these gods as the essence and power of earth and water."[255]

The Austro-Hungarians have a man in the moon who is a sort of aquarius. Grimm says: "Water, an essential part of the Norse myth, is wanting in the story of the man with the thorn bush, but it reappears in the Carniolan story cited in Bretano's Libussa (p. 421): the man in the moon is called Kotar, he makes her grow by pouring water."[256] The Scandinavian legend, distilled into Jack and Jill, is, as we have seen, an embodiment of early European belief that the ebb and flow of the tides were dependent upon the motions and mutations of the moon.

We find the same notion prevailing in the western hemisphere. "As the MOON is associated with the dampness and dews of night, an ancient and widespread myth identified her with the goddess of water. Moreover, in spite of the expostulations of the learned, the common people the world over persist in attributing to her a marked influence on the rains. Whether false or true, this familiar opinion is of great antiquity, and was decidedly approved by the Indians, who were all, in the words of an old author, 'great observers of the weather by the moon.' They looked upon her, not only as forewarning them by her appearance of the approach of rains and fogs, but as being their actual cause. Isis, her Egyptian title, literally means moisture; Ataensic, whom the Hurons said was the moon, is derived from the word for water; and Citatli and Atl, moon and water, are constantly confounded in Aztec theology."[257] One of the gods of the Dakotahs was "Unk-ta-he (god of the water). The Dakotahs say that this god and its associates are seen in their dreams. It is the master-spirit of all their juggling and superstitious belief, From it the medicine men obtain their supernatural powers, and a great part of their religion springs from this god."[258] Brinton also says of this large Indian nation, "that Muktahe, spirit of water, is the master of dreams and witchcraft, is the belief of the Dakotahs."[259] We know that the Dakotahs worshipped the moon, and therefore see no difficulty in identifying that divinity with their god of dreams and water. "In the legend of the Muyscas it is Chia, the moon, who was also goddess of water and flooded the earth out of spite."[260] In this myth the moon is a malevolent deity, and water, usually a symbol of life, becomes an agency of death. Reactions are constantly occurring in the myth-making process. The god is male or female, good or evil, angry or amiable, according to the season or climate, the aspect of nature or the mood of the people. "In hot countries," says Sir John Lubbock, "the sun is generally regarded as an evil, and in cold as a beneficent being."[261] We are willing to accept this, with allowance. There is little question that taking men as a whole they are mainly optimistic in their judgments respecting the gifts of earth and the glories of heaven. Mr. Brinton, in reference to the imagined destructiveness of the water deity, writes: "Another reaction in the mythological laboratory is here disclosed. As the good qualities of water were attributed to the goddess of night, sleep, and death, so her malevolent traits were in turn reflected back on this element. Taking, however, American religions as a whole, water is far more frequently represented as producing beneficent effects than the reverse."[262]

"The time of full moon was chosen both in Mexico and Peru

to celebrate the festival of the deities of water, the patrons of agriculture, and very generally the ceremonies connected with the crops were regulated by her phases. The Nicaraguans said that the god of rains, Quiateot, rose in the east, thus hinting how this connection originated."[263] "The Muyscas of the high plains of Bogota were once, they said, savages without agriculture, religion, or law; but there came to them from the east an old and bearded man, Bochica, the child of the sun, and he taught them to till the fields, to clothe themselves, to worship the gods, to become a nation. But Bochica had a wicked, beautiful wife, Huythaca, who loved to spite and spoil her husband's work; and she it was who made the river swell till the land was covered by a flood, and but a few of mankind escaped upon the mountain tops. Then Bochica was wroth, and he drove the wicked Huythaca from the earth, and made her the moon, for there had been no moon before; and he cleft the rocks and made the mighty cataract of Tequendama, to let the deluge flow away. Then, when the land was dry, he gave to the remnant of mankind the year and its periodic sacrifices, and the worship of the sun. Now the people who told this myth had not forgotten, what indeed we might guess without their help, that Bochica was himself Zuhé, the sun, and Huytheca, the sun's wife, the moon."[264] This interesting and instructive legend, to which we alluded before in a brief quotation from Mr. Brinton, is worthy of reproduction in its fuller form, and fitly concludes our moon mythology and worship, as it presents a synoptical view of the chief points to which our attention has been turned. It shows us primitive or primeval man, the dawn of civilization, the daybreak of religion, the upgrowth of national life. In its solar husband and lunar wife it embraces that anthropomorphism and sexuality which we think have been and still are the principal factors in the production of legendary and religious impersonations. It includes that dualism which is one of man's oldest attempts to account for the opposition of good and evil. And finally it predicts a new humanity, springing from a remnant of the old; and a progress of brighter years, when, the deluge having disappeared, the dry land shall be fruitful in every good; when men shall worship the Father of lights, and "God shall be all in all."*

* For further information on the universality of moon-worship, see *The Ceremonies and Religious Customs of the Various Nations of the Known World*, by Bernard Picart. London: 1734, folio, vol. iii.

MOON SUPERSTITIONS

Introduction

Superstition may be defined as an extravagance of faith and fear: not what Ecclesiastes calls being "righteous overmuch," but religious reverence in excess. Some etymologists say that the word originally meant a "*standing* still *over* or by a thing" in fear, wonder, or dread.[265] Brewer's definition is rather more classical: "That which survives when its companions are dead (Latin, *supersto*). Those who escaped in battle were called *superstitës*. Superstition is that religion which remains when real religion is dead; that fear and awe and worship paid to the religious impression which survives in the mind when correct notions of Deity no longer exist."[266] Hooker says that superstition "is always joined with a wrong opinion touching things divine. Superstition is, when things are either abhorred or observed with a zealous or fearful, but erroneous relation to God. By means whereof the superstitious do sometimes serve, though the true God, yet with needless offices, and defraud Him of duties necessary; sometimes load others than Him with such honours as properly are His."[267] A Bampton Lecturer on this subject says: "Superstition is an *unreasonable belief* of that which is mistaken for truth concerning the nature of God and the invisible world, our relations to these unseen objects, and the duties which spring out of those relations."[268]

We may next briefly inquire into the origin of the thing, which, of course, is older than the word. Burton will help us to an easy answer. He tells us that "the *primum mobile*, and first mover of all superstition, is the devil, that great enemy of mankind, the principal agent, who in a thousand several shapes, after divers fashions, with several engines, illusions, and by several names, hath deceived the inhabitants of the earth, in several places and countries, still rejoicing at their falls."[269] Verily this protean, omnipresent, and malignant devil has proved himself a great convenience! He has been the scapegoat upon whom we have laid the responsibility of all our mortal woe: and now we learn that to his infernal influence we are indebted for our ignorance and superstition. Henceforth, when we are at our wit's end, we may apostrophize the difficulty, and exclaim, "O thou invisible spirit, if thou hast no name to be known by, let us call thee devil!" We hesitate to spoil this serviceable illusion: for as we have known some good people, of a sort, who would be distressed to find that there was no hell to burn up the opponents of their orthodoxy; we fear lest many would be disappointed if they found out that the infernal spirit was not at the bottom of our abysmal ignorance. But we will give even the devil his due. We are not like Sir William Brown, who "could never bring himself heartily to hate the devil." We can, wherever we find him; but we think it only honest to father our own mental deficiencies, as well as our moral delinquencies, and instead of seeking a substitute to use the available remedy. "To err is human"; and it is in humanity itself that we shall discover the source of superstition. We are the descendants of ancestors who were the children of the world, and we were ourselves children not so long ago. Childhood is the age of fancy and fiction; of sensitiveness to outer influences; of impressions of things as they seem, not as they are. When we become men we put away childish things; and in the manhood of our race we shall banish many of the idols and ideas which please us while we grow. Darwin has told us that our "judgment will not rarely err from ignorance and weak powers of reasoning. Hence the strangest customs and superstitions, in complete opposition to the true welfare and happiness of mankind, have become all-powerful throughout the world. How so many absurd rules of conduct, as well as so many absurd religious beliefs, have originated, we do not know; nor how it is that they have become, in all quarters of the world, so deeply impressed on the mind of men; but it is worthy of remark that a belief constantly inculcated during the early years of life, whilst the brain is impressible, appears to acquire almost the nature of an instinct; and the very essence of an instinct is that it is followed independently of reason."[270]

But if superstition be the result of imperfection, there is no gainsaying the fact that it is productive of infinite evil; and on this account it has been attributed to a diabolical paternity. Bacon even affirms that "it were better to have no opinion of God at all, than such an opinion as is unworthy of Him; for the one is unbelief, the other is contumely: and certainly superstition is the reproach of the Deity."[271] Most heartily do we hold with Dr. Thomas Browne: "It is not enough to believe in God as an irresistible power that presides over the universe; for this a malignant demon might be. It is necessary for our devout happiness that we should believe in Him as that pure and gracious Being who is the encourager of our virtues and the comforter of our sorrows.

Quantum religio potuit suadere malorum,

exclaims the Epicurean poet, in thinking of the evils which superstition, characterized by that ambiguous name, had produced; and where a fierce or gloomy superstition has usurped the influence which religion graciously exercises only for purposes of benevolence to man, whom she makes happy with a present enjoyment, by the very expression of devout gratitude for happiness already enjoyed, it would not be easy to estimate the amount of positive misery which must result from the mere contemplation of a tyrant in the heavens, and of a creation subject to his cruelty and caprice."[272] The above quoted line from Lucretius—To such evils could religion persuade!—is more than the exclamation of righteous indignation against the sacrifice of Iphigenia by her father, Agamemnon, at the bidding of a priest, to propitiate a goddess. It is still further applicable to the long chain of outrageous wrongs which have been inflicted upon the innocent at the instigation of a stupid and savage fanaticism. What is worst of all, much of this bloodthirsty religion has claimed a commission from the God of love, and performed its detestable deeds in the insulted name of that "soft, meek, patient, humble, tranquil spirit," whom the loftiest and best of men delight to adore as the Prince of peace. No wonder that Voltaire cried out, "Christian religion, behold thy consequences!" if he could calculate that ten million lives had been immolated on the altar of a spurious Christianity. One hundred thousand were slain in the Bartholomew massacre alone. Righteousness, peace, and love were not the monster which Voltaire laboured to crush: he was most intensely incensed against the blind and bigoted priesthood, against the malicious and murderous servants who ate the bread of a holy and harmless Master, against "their intolerance of light and hatred of knowledge, their fierce yet profoundly contemptible struggles with one another, the scandals of their casuistry, their besotted cruelty."[273] We have been betrayed into speaking thus strongly of the extreme lengths to which superstition will carry those who yield themselves to its ruthless tyranny. But perhaps we have not gone far from our subject, after all; for the innocent Iphigenia, whose doom kindled our ire, was sacrificed to the goddess of the moon.

Lunar Fancies

There are a few phosphorescent fancies about the moon, like *ignes fatui*,

> "Dancing in murky night o'er fen and lake,"

which we may dispose of in a section by themselves. Those of them that are mythical are too evanescent to become full-grown myths; and those which are religious are too volatile to remain in the solution or salt of any bottled creed. Like the wandering lights of the Russians, answering to our will-o'-the-wisp, they are the souls of still-born children. There is, for example, the insubstantial and formless but pleasing conception of the Indian Veda. In the Râmâyanam the moon is a good fairy, who in giving light in the night assumes a benignant aspect and succours the dawn. In the Vedic hymn, Râkâ, the full moon, is exhorted to sew the work with a needle which cannot be broken. Here the moon is personified as preparing during the night her luminous garments, one for the evening, the other for the morning, the one lunar and of silver, the other solar and of gold.[274] Another notion, equally airy but more religious, has sprung up in Christian times and in Catholic countries. It is that heathen fancy which connects the moon with the Virgin Mary. Abundant evidence of this association in the minds of Roman Catholics is furnished by the style of the ornaments which crowd the continental churches. One of the most conspicuous is the sun and moon in conjunction, precisely as they are represented on Babylonian and Grecian coins; and the identification of the Virgin and her Child with the moon any Roman Catholic cathedral will show.[275] The *Roman Missal* will present to any reader "Sancta Maria, coeli Regina, et mundi Domina"; the *Glories of Mary* will exhibit her as the omnipotent mother, Queen of the Universe; and Ecclesiastical History will declare how, as early as the close of the fourth century, the women who were called Collyridians worshipped her "as a goddess, and judged it necessary to appease her anger, and seek her favour and protection, by libations, sacrifices, and oblations of cakes (*collyridae*)."[276] This is but a repetition of the women kneading dough to make cakes to the queen of heaven, as recorded by Jeremiah; and proves that the relative position occupied by Astarte in company with Baal, Juno with Jupiter, Doorga with Brahma, and Ma-tsoo-po with Boodh, is that occupied by Mary with God. Nay more, she is "Mater Creatoris" and "Dei Genetrix": Mother of the Creator, Mother of God. Having thus been enthroned in the position in the universal pantheon which was once occupied by the moon, what wonder that the ignorant devotee should see her in that orb, especially as the sun, moon, and stars of the Apocalypse are her chief symbols. Southey has recorded a good illustration of this superstitious fancy. "A fine circumstance occurred in the shipwreck of the *Santiago*, 1585. The ship struck in the night; the wretched crew had been confessing, singing litanies, etc., and this they continued till, about two hours before break of day, the moon arose beautiful and exceeding bright; and forasmuch as till that time they had been in such darkness that they could scarcely sec one another when close at hand, such was the stir among them at beholding the brightness and glory of that orb, that most part of the crew began to lift up their voices, and with tears, cries, and groans called upon Our Lady, saying they saw her in the moon."[277]

The preceding fancies would produce upon the poetic and religious sense only an agreeable effect. Other hallucinations have wrought effects of an opposite kind. The face in the moon does not always wear an amiable aspect, and it is not unnatural that those who have been taught to believe in angry gods and frowning providences should see the caricatures of their false teachers reproduced in the heavens above and in the earth beneath. We are reminded here of the magic mirror mentioned by Bayle. There is a trick, invented by Pythagoras, which is performed in the following manner. The moon being at the full, some one writes with blood on a looking-glass anything he has a mind to; and having given notice of it to another person, he stands behind that other and turns towards the moon the letters written in the glass. The other looking fixedly on the shining orb reads in it all that is written on the mirror as if it were written on the moon.[278] This is precisely the *modus operandi* by which the knavish have imposed upon the foolish in all ages. The manipulator of the doctrine stands behind his credulous disciple, writing out of sight his invented science or theology, and writing too often with the blood of some innocent victim. The poor patient student is meanwhile gazing on the moon in dreamy devotion; until as the writing on the mirror is read with solemn intonation, it all appears before his moon-struck gaze as a heavenly revelation. Woe to the truth-loving critic who breaks the enchantment and the mirror, crying out in the vernacular tongue, Your mysteries are myths, your writings are frauds; and the fair moon is innocent of the lying imposition!

To multitudes the moon has always been an object of terror and dread. Not only is it a supramundane and magnified man—that it will always be while its spots are so anthropoid, and man himself is so anthropomorphic—but it has ever been, and still is, a being of maleficent and misanthropic disposition. As Mr. Tylor says, "When the Aleutians thought that if any one gave offence to the moon, he would fling down stones on the offender and kill him; or when the moon came down to an Indian squaw, appearing in the form of a beautiful woman with a child in her arms, and demanding an offering of tobacco and fur-robes: what conceptions of personal life could be more distinct than these?"[279] Personal and distinct, indeed, but far from pleasant. Another author tells us that "in some parts of Scotland to point at the stars or to do aught that might be considered an indignity in the face of the sun or moon, is still to be dreaded and avoided; so also it was not long since, probably still is, in Devonshire and Cornwall. The Jews seem to have been equally superstitious on this point (Jer. viii. 1, 2), and the Persians believed leprosy to be an infliction on those who had committed some offence against the sun."[280] Southey supplies us with an illustration of the moon in a fit of dudgeon. He is describing the sufferings of poor Hans Stade, when he was caught by the Tupinambas and expected that he was about to die. "The moon was up, and fixing his eyes upon her, he silently besought God to vouchsafe him a happy termination of these sufferings. Yeppipo Wasu, who was one of the chiefs of the horde, and as such had convoked the meeting, seeing how earnestly he kept gazing upwards, asked him what he was looking at. Hans had ceased from praying, and was observing the man in the moon, and fancying that he looked angry; his mind was broken down by continual terror, and he says it seemed to him at that moment as if he were hated by God, and by all things which God had created. The question only half roused him from this phantasy, and he answered, it was plain that the moon was angry. The savage asked whom she was angry with, and then Hans, as if he had recollected himself, replied that she was looking at his dwelling. This enraged him, and Hans found it prudent to say that perhaps her eyes were turned so wrathfully upon the Carios; in which opinion the chief assented, and wished she might destroy them all."[281] Some such superstitious fear must have furnished the warp into which the following Icelandic story was woven. "There was once a sheep-stealer who sat down in a lonely place, with a leg of mutton in his hand, in order to feast upon it, for he had just stolen it. The moon shone bright and clear, not a single cloud being there in heaven to hide her. While enjoying his gay feast, the impudent thief cut a piece off the meat, and, putting it on the point of his knife, accosted the moon with these godless words:—

'O moon, wilt thou
On thy mouth now
This dainty bit of mutton-meat?'

Then a voice came from the heavens, saying:—

'Wouldst thou, thief, like
Thy cheek to strike
This fair key, scorching-red with heat?'

At the same moment, a red-hot key fell from the sky on to the cheek of the thief, burning on it a mark which he carried with him ever afterwards. Hence arose the custom in ancient times of branding or marking thieves."[282] The moral influence of this tale is excellent, and has the cordial admiration of all who hate robbery and effrontery: at the same time it exhibits the moon as an irascible body, with which no liberty may be taken. In short, it is an object of superstitious awe.

One other lunar fancy, born and bred in fear, is connected with the abominable superstition of witchcraft. Abominable, unquestionably, the evil was; but justice compels us to add that the remedy of relentless and ruthless persecution with which it was sought to remove the pest was a reign of abhorrent and atrocious cruelty. Into the question itself we dare not enter, lest we should be ourselves bewitched. We know that divination by supposed supernatural agency existed among the Hebrews, that magical incantations were practised among the Greeks and Romans, and that more modern witchcraft has been contemporaneous with the progress of Christianity. But we must dismiss the subject in one borrowed sentence. "The main source from which we derived this superstition is the East, and traditions and facts incorporated in our religion. There were only wanted the ferment of thought of the fifteenth century, the energy, ignorance, enthusiasm, and faith of those days, and the papal denunciation of witchcraft by the bull of Innocent the Eighth, in 1459, to give fury to the delusion. And from this time, for three centuries, the flames at which more than a hundred thousand victims perished cast a lurid light over Europe."[283] The singular notion, which we wish to present, is the ancient belief that witches could control the moon. In the *Clouds* of Aristophanes, Strepsiades tells Socrates that he has "a notion calculated to deprive of interest"; which is as follows:—

"*Str.* If I were to buy a Thessalian witch, and draw down the moon by night, then shut her up in a round helmet-case, like a mirror, and then keep watching her—"

"*Soc.* What good would that do you, then?"

"*Str.* What? If the moon were not to rise any more anywhere, I should not pay the interest."

"*Soc*. Because what?"

"*Str.* Because the money is lent by the month."[284]

Shakespeare alludes to this, where Prospero says, "His mother was a witch, and one so strong that could control the moon" (*Tempest*, Act v.).

If the witch's broom, on whose stick she rode to the moon, be a type of the wind, we may guess how the fancy grew up that the airy creation could control those atmospheric vapours on which the light and humidity of the night were supposed to depend.[285]

Lunar Eclipses

All round the globe, from time immemorial, those periodic phenomena known as solar and lunar eclipses have been occasions of mental disquietude and superstitious alarm. Though now regarded as perfectly natural and regular, they have seemed so preternatural and irregular to the unscientific eye that we cannot wonder at the consternation which they have caused. And it must be confessed that a total obscuration of the sun in the middle of the day casts such a gloom over the earth that men not usually timid are still excusable if during the parenthesis they feel a temporary uneasiness, and are relieved when the ruler of the day emerges from his dark chamber, apparently rejoicing to renew his race. An eclipse of the moon, though less awe-inspiring, is nevertheless sufficiently so to awaken in the superstitious brain fearful forebodings of impending calamity. Science may demonstrate that there is nothing abnormal in these occurrences, but to the seeker after signs it wilt be throwing words away; for, as Lord Kames says, "Superstitious eyes are never opened by instruction."

We will now produce a number of testimonies to show how these lunar eclipses have been viewed among the various races of the earth in ancient and modern times. The Chaldaeans were careful observers of eclipses, and Berosus believed that when the moon was obscured she turned to us her dark side. Anaximenes said that her mouth was stopped. Plato, Aristotle, the Stoics, and the Mathematicians said that she fell into conjunction with the bright sun. Anaxagoras of Clazomenae (born B.C. 499) was the first to explain the eclipse of the moon as caused by the shadow of the earth cast by the sun. But he was as one born out of due time. We are all familiar with the use made by students of unfulfilled prophecy of every extraordinary occurrence in nature, such as the sudden appearance of a comet, an earthquake, an eclipse, etc. We know how mysteriously they interpret those simple passages in the Bible about the sun being darkened and the moon being turned into blood. If they were not wilfully blind, such facts as are established by the following quotations would open their eyes to the errors in their exegesis. At any rate, they would find their theories anticipated in nearly every particular by those very heathen whom they are wont to pity as so benighted and hopelessly lost.

Grimm writes: "One of the most terrible phenomena to heathens was an *eclipse* of the sun or moon, which they associated with a destruction of all things and the end of the world. I may safely assume that the same superstitious notions and practices attend eclipses among nations ancient and modern. The Indian belief is that a serpent eats up the sun and moon when they are eclipsed, or a demon devours them. To this day the Hindoos consider that a giant lays hold of the luminaries and tries to swallow them. The Chinese call the solar eclipse zhishi (solis devoratio), the lunar yueshi (lunae devoratio), and ascribe them both to the machinations of a dragon. Nearly all the populations of Northern Asia hold the same opinion. The Finns of Europe, the Lithuanians, and the Moors in Africa, have a similar belief."[286] Flammarion says: "Among the ancient nations people used to come to the assistance of the moon, by making a confused noise with all kinds of instruments, when it was eclipsed. It is even done now in Persia and some parts of China, where they fancy that the moon is fighting with a great dragon, and they think the noise will make him loose his hold and take to flight. Among the East Indians they have the same belief that when the sun and the moon are eclipsed, a dragon is seizing them, and astronomers who go there to observe eclipses are troubled by the fears of their native attendants, and by their endeavours to get into the water as the best place under the circumstances. In America the idea is that the sun and moon are tired when they are eclipsed. But the more refined Greeks believed for a long time that the moon was bewitched, and that the magicians made it descend from heaven to put into the herbs a certain maleficent froth. Perhaps the idea of the dragon arose from the ancient custom of calling the places in the heavens at which the eclipses of the moon took place the head and tail of the dragon."[287]

Sir Edward Sherburne, in his "Annotations upon the *Medea*," quaintly says: "Of the beating of kettles, basons, and other brazen vessels used by the ancients when the moone was eclipsed (which they did to drown the charms of witches, that the moon might not hear them, and so be drawne from her spheare as they suppos'd), I shall not need to speake, being a thing so generally knowne, a custom continued among the Turks to this day; yet I cannot but adde, and wonder at, what Joseph Scaliger, in his 'Annotations upon Manilius,' reports out of Bonincontrius, an ancient commentator upon the same poet, who affirms that in a town of Italy where he lived (within these two centuries of yeares), he saw the same piece of paganisme acted upon the like occasion."[288] Another, and more recent writer, also says of these eclipses: "The Chinese imagine them to be caused by great dragons trying to devour the sun and moon, and beat drums and brass kettles to make the monsters give up their prey. Some of the tribes of American Indians speak of the moon as hunted by huge dogs, catching and tearing her till her soft light is reddened and put out by the blood flowing from her wounds. To this day in India the native beats his gong, as the moon passes across the sun's face, and it is not so very long ago that in Europe both eclipses and rushing comets were thought to show that troubles were near."[289] Respecting China, a modern traveller speaks in not very complimentary language. "If there is on the earth a nation absorbed by the affairs of this world and who trouble themselves little about what passes among the heavenly bodies, it is assuredly the Chinese. The most erudite among them just know of the existence of astronomy, or, as they call it, *tienwen*—'celestial literature.' But they are ignorant of the simplest principles of the science, and those who regard an eclipse as a natural phenomenon, instead of a dragon who is seeking to devour the sun and moon, are enlightened indeed."[290] This statement ought to be taken with more than one *granum salis*, especially as Mrs. Somerville assures us that the Chinese had made advances in the science of astronomy 1,100 years before the Christian era, and also adds: "Their whole chronology is founded on the observation of eclipses, which prove the existence of that empire for more than 4,700 years."[291] With this discount the charge against Chinese ignorance may be passed. "A Mongolian myth makes out that the gods determined to punish Arakho for his misdeeds, but he hid so effectually that no one could find out his lurking-place. They therefore asked the *sun*, who gave an unsatisfactory answer; but when they asked the *moon*, she disclosed his whereabouts. So Arakho was dragged forth and chastised; in revenge of which he *pursues both sun and moon*, and whenever he comes to hand grips with one of them, *an eclipse occurs*. To help the lights of heaven in their sad plight, a *tremendous uproar* is made with musical and other instruments, till Arakho is scared away."[292] "Referring to the Shoo, Pt. III., Bk. IV., parag. 4, we find this sentence: 'On the first day of the last month of autumn the sun and moon did not meet harmoniously in Fang.'"[293] In less euphemistic phrase, the sun and moon were *crossed*.

Dr. Wells Williams describes an interesting scene. "In the middle of the sixth moon lanterns are hung from the top of a pole placed on the highest part of the house. A single small lantern is deemed sufficient, but if the night be calm, a greater display is made by some householders, and especially in boats, by exhibiting coloured glass lamps arranged in various ways. The illumination of a city like Canton, when seen from a high spot, is made still more brilliant by the moving boats on the river. On one of these festivals at Canton, an almost total eclipse of the moon called out the entire population, each one carrying something with which to make a noise, kettles, pans, sticks, drums, gongs, guns, crackers, and what not to frighten away the dragon of the sky from his hideous feast. The advancing shadow gradually caused the myriads of lanterns to show more and more distinctly, and started a still increasing clamour, till the darkness and the noise were both at their climax. Silence gradually resumed its sway as the moon recovered her fulness."[294] On another page Dr. Williams tells us that "some clouds having on one occasion covered the sky, so that an eclipse could not be seen, the courtiers joyfully repaired to the emperor to felicitate him that Heaven, touched by his virtues, had spared him the pain of witnessing the 'eating of the sun.'"[295] The following passage from Doolittle's work on the Chinese is sufficiently interesting to be given without abridgment: "It is a part of the official duties of mandarins to 'save the sun and moon when eclipsed.' Prospective eclipses are never noticed in the Imperial Calendar, published originally at Peking, and republished in the provinces. The imperial astronomers at the capital, a considerable time previous to a visible eclipse, inform the Board of Rites of its month, day, and hour. These officers send this intelligence to the viceroys or governors of the eighteen provinces of the empire. These, in turn, communicate the information to all the principal subordinate officers in the provinces of the civil and the military grade. The officers make arrangements to save the moon or the sun at the appointed time. On the day of the eclipse, or on the day preceding it, some of them put up a written notice in or near their yamuns, for the information of the public.

"The Chinese generally have no rational idea of the cause of eclipses. The common explanation is that the sun or the moon has experienced some disaster. Some even affirm that the object eclipsed is being devoured by an immense ravenous monster. This is the most popular sentiment in Fuhchau in regard to the procuring cause of eclipses. All look upon the object eclipsed with wonder. Many are filled with apprehension and terror. Some of the common people, as well as mandarins generally, enter upon some course of action, the express object of which is to save the luminary from its dire calamity, or to rescue it from the jaws of its greedy enemy. Mandarins must act officially, and in virtue of their being officers of government. Neither they nor the people seem to regard the immense distance of the celestial object as at all interfering with the success of their efforts. The various obstacles which ought apparently to deter them from attempting to save the object eclipsed do not seem to have occurred to them at all, or, if they have occurred, do not appear to be sufficient to cause them to desist from prosecuting their laudable endeavours. The high mandarins procure the aid of priests of the Taoist sect at their yamuns. These place an incense censer and two large candlesticks for holding red candles or tapers on a table in the principal reception room of the mandarin, or in the open space in front of it under the open heavens.

"At the commencement of the eclipse the tapers are lighted, and soon after the mandarin enters, dressed in his official robes. Taking some sticks of lighted incense in both hands, he makes his obeisance before or facing the table, raising and depressing the incense two or three times, according to the established fashion, before it is placed in the censer. Or sometimes the incense is lighted and put in the censer by one of the priests employed. The officer proceeds to perform the high ceremony of kneeling down three times, and knocking his head on the ground nine times. After this he rises from his knees. Large gongs and drums near by are now beaten as loudly as possible. The priests begin to march slowly around the tables, reciting formulas, etc., which marching they keep up, with more or less intermissions, until the eclipse has passed off.

"A uniform result always follows these official efforts to save the sun and the moon. *They are invariably successful.* There is not a single instance recorded in the annals of the empire when the measures prescribed in instructions from the emperor's astronomers at Peking, and correctly carried out in the provinces by the mandarins, have not resulted in a complete rescue of the object eclipsed. Doubtless the vast majority of the common people in China believe that the burning of tapers and incense, the prostration of the mandarins, the beating of the gongs and drums, and the recitations on the part of the priests, are signally efficacious in driving away the voracious monster. They observe that the sun or the moon does not seem to be permanently injured by the attacks of its celestial enemy, although a half or nearly the whole appeared to have been swallowed up. This happy result is doubtless viewed with much complacency by the parties engaged to bring it about. The lower classes generally leave the saving of the sun or the moon, when eclipsed, to their mandarins, as it is a part of their official business. Some of the people occasionally beat in their houses a winnowing instrument, made of bamboo splints, on the occasion of an eclipse. This gives out a loud noise. Some venture to assert that the din of this instrument penetrates the clouds as high as the very temple of Heaven itself! The sailors connected with junks at this place, on the recurrence of a lunar eclipse, always contribute their aid to rescue the moon by beating their gongs in a most deafening manner.

"Without doubt, most of the mandarins understand the real occasion of eclipses, or, at least, they have the sense to perceive that nothing which they can do will have any effect upon the object eclipsed, or the cause which produces the phenomenon; but they have no optional course in regard to the matter. They must comply with established custom, and with the understood will of their superiors. The imperial astronomers, having been taught the principles of astronomy and the causes which produce eclipses by the Roman Catholic missionaries a long while since, of course know that the common sentiments on the subject are as absurd as the common customs relating to it are useless. But the emperor and his cabinet cling to ancient practices, notwithstanding the clearest evidences of their false and irrational character."[296]

Mr. Herbert Giles accounts for this Chinese obtuseness, or, as some would have it, opacity, in much the same way. Under the head of *Natural Phenomena*, he writes: "It is a question of more than ordinary interest to those who regard the Chinese people as a worthy object of study, What are the speculations of the working and uneducated classes concerning such natural phenomena as it is quite impossible for them to ignore? Their theory of eclipses is well known, foreign ears being periodically stunned by the gonging of an excited crowd of natives, who are endeavouring with hideous noises to prevent some imaginary dog of colossal proportions from banqueting, as the case may be, upon the sun or moon. At such laughable exhibitions of native ignorance it will be observed there is always a fair sprinkling of well-to-do, educated persons, who not only ought to know better themselves, but should be making some effort to enlighten their less fortunate countrymen instead of joining in the din. Such a hold, however, has superstition on the minds of the best informed in a Chinese community, that under the influence of any real or supposed danger, philosophy and Confucius are scattered to the four winds of heaven, and the proudest disciple of the master proves himself after all but a man."[297] No doubt Mr. Doolittle and Mr. Giles are both right: custom and superstition form a twisted rope which pinions the popular mind. But there is yet another strand to be mentioned which makes the bond a threefold cord which it will take some time to break. *Prescriptive right* requires that the official or cultured class in China, answering to the clerical caste elsewhere, should keep the other classes in ignorance; because, if science and religion are fellow-helpers, science and superstition can never dwell together, and the downfall of superstition in China would be the destruction of imperial despotism and magisterial tyranny. "Sirs, ye know that by this craft we have our wealth. But this Paul says that they be no gods, which are made with hands: so that our craft is in danger to be set at nought. Great is Diana of the Ephesians!" The mandarins know why they encourage the mechanics and merchants to save the moon.

We once met a good story in reading one of Jean Astruc's medical works. "Theodore de Henry, of Paris, coming one time into the church of St. Dionis, he fell prostrate at the foot of the statue of Charles the Eighth, as in a sudden fit of devotion. When being told by one of the monks that was not the image of any saint, he replied, he was not ignorant of that, but was willing to pay a grateful acknowledgment to the memory of that prince who had brought the *Morbus Gallicus* into France, by which he had made his own fortune." Herein lies the secret of half of the hypocrisy of the world. Thank God! the world moves; and the millennium of truth is at hand.

The literature of China is, happily, not all linsey-woolsey. The following sample is of the finest silk, worthy to adorn the purest saint.

"Ming Ti of the House of Wei.

"Reigned 227-239 A.D.

"*On an Eclipse.—A Rescript.* WE have heard that if a sovereign is remiss in government, Heaven terrifies him by calamities and strange portents. These are divine reprimands sent to recall him to a sense of duty. Thus, partial eclipses of the sun and moon are manifest warnings that the rod of empire is not wielded aright. Ever since WE ascended the throne, OUR inability to continue the glorious traditions of our departed ancestors and carry on the great work of civilization, has now culminated in a warning message from on high. It therefore behoves Us to issue commands for personal reformation, in order to avert the impending calamity.

"But the relations of Heaven with Man are those of a father and son; and a father about to chastise his son would not be deterred were the latter to present him with a dish of meat. WE do not therefore consider it part of OUR duty to act in accordance with certain memorials advising that the prime minister and chief astronomer be instructed to offer up sacrifices on this occasion. Do ye, governors of districts and other high officers of State, seek rather to rectify your own hearts; and if any one can devise means to make up for OUR shortcomings, let him submit his proposals to the Throne."[298]

The writer of that was "not far from the kingdom of God."

Father Borri, in his account of Cochin China, describes the effect of a lunar eclipse upon several scholars in the city of Nuoecman in the province of Pulucambi. "I showed them that the circle of the moon, on that side the eclipse began, was not so perfect as it should be, and soon after all the moon being darkened, they perceived the truth of my prediction. The commander and all of them being astonished, presently sent to give notice of it to all the ward, and spread the news of the eclipse throughout the city, that every man might go out to make the usual noise in favour of the moon; giving out everywhere that there were no such men as the fathers, whose doctrine and books could not fail being true, since they had so exactly foretold the eclipse, which their learned men had taken no notice of; and therefore, in performance of his promise, the commander with all his family became Christians, as did many more of his ward, with some of the most learned men of the city and others of note."[299] In no unkind spirit we cannot refrain from noticing, what will strike every reader, how ready divines of all denominations are to turn the teachings of science to their own account in the propagation of their faith. It would have been seemlier for theologians in all ages, if their attitude towards physical inquirers had been less hostile; they would then have made converts through eclipses with a better grace. They would, moreover, have prevented the alienation of many of their truest friends.

Captain Beeckman gives an amusing story of an eclipse in Cantongee, in the island of Borneo, on the 10th of November, 1714. "We sat very merry till about eight at night, when, preparing to go to bed, we heard all on a sudden a most terrible outcry, mixed with squealing, halloing, whooping, firing of guns, ringing and clattering of gongs or brass pans, that we were greatly startled, imagining nothing less but that the city was surprised by the rebels. I ran immediately to the door, where I found my old fat landlord roaring and whooping like a man raving mad. This increased my astonishment, and the noise was so great that I could neither be heard, nor get an answer to know what the matter was. At last I cried as loud as possibly I could to the old man to know the reason of this sad confusion and outcry, who in a great fright pointed up to the heavens, and said, '*Look there; see, the devil is eating up the moon*!' I was very glad to hear that there was no other cause of their fright but their own ignorance. It was only a great eclipse of the moon. I smiled, and told him that there was no danger; that in a little while the moon would be as well as ever. Whereupon, catching fast hold of my sleeve, as I was returning to bed, he asked me if I was sure on't (for they take us white men to be very wise in those matters). I assured him I was, and that we always knew many years before when such a thing would happen; that it proceeded from a natural cause, according to the course and motion of the sun and moon, and that the devil had no hand in it. After the eclipse was over, the old man, being not a little rejoiced, took me in."[300] Another writer speaks of the East India Islands in general. "There is to this day hardly a country of the Archipelago in which the ceremony of frightening the supposed monster from his attack on the luminary is not performed. This consists in shouting, in striking gongs, but, above all, in striking their stampers against the sides of the wooden mortars which are used by the villagers in husking their corn."[301] That the Indians of the continent regard the phenomena in question with more than ordinary interest is evinced by their resorting in large numbers to Benares, the ancient seat of brahminical learning and religion, on every occasion of an eclipse of the moon. Lord Kames reminds us that among the Greeks "an eclipse being held a prognostic given by the gods of some grievous calamity, Anaxagoras was accused of atheism for attempting to explain the eclipse of the moon by natural causes: he was thrown into prison, and with difficulty was relieved by the influence of Pericles. Protagoras was banished Athens for maintaining the same doctrine."[302]

Thucydides tells us that an eclipse of the moon delayed the departure of the expedition against the Syracusans. "The preparations were made, and they were on the point of sailing, when the moon, being just then at the full, was eclipsed. The mass of the army was greatly moved, and called upon the generals to remain. Nicias himself, who was too much under the influence of divination and omens, refused even to discuss the question of their removal until they had remained thrice nine days, as the soothsayers prescribed. This was the reason why the departure of the Athenians was finally delayed."[303]

"At any eclipse of the moone, the Romanes would take their brazen pots and pannes, and beat them, lifting up many torches and linckes lighted, and firebrandes into the aire, thinking by these superstitious meanes to reclaime the moone to her light."[304]

The Constantinople Messenger of December 23rd, 1880, contains the following:—"Mgr. Mamarbasci, who represents the Syrian Patriarch at the Porte, and who resides in St. Peter's Monastery in Galata, underwent a singular experience on the evening of the last eclipse of the moon. Hearing a great noise outside of the firing of revolvers and pistols, he opened his window to see what could be the cause of so much waste of powder. Being a native of Aleppo, he was at no loss to understand the cause of the disturbance as soon as he cast his eye on the heavens, and he therefore immediately withdrew his head from the window again. Hardly had he done so, however, ere a ball smashed the glass into a thousand pieces. Rising from the seat into which he had but just sat down, he perceived a conical ball on the floor of his room, which there is every reason to believe would have killed him on the spot had he remained a moment longer on the spot he had just quitted. From the yard of the mosque of Arab-Djami, which is in front of the prelate's window, the bullet had, it appears, been fired with the intention of frightening the dragon or bear which, according to oriental superstition, lies in wait to devour the moon at its eclipse. It is a fortunate circumstance that the Syrian ecclesiastic escaped scathless from the snares laid to destroy the celestial dragon."[305]

In the *Edda*, an ancient collection of Scandinavian poetry, embodying the national mythology, Managarmer is the monster who sometimes swallows up the moon, and stains the heaven and the air with blood. "Here," says M. Mallett, "we have the cause of eclipses; and it is upon this very ancient opinion that the general practice is founded, of making noises at that time, to fright away the monster, who would otherwise devour the two great luminaries."[306] Of the Germans, Grimm says:—"In a lighted candle, if a piece of the wick gets half detached and makes it burn away too fast, they say 'a *wolf* (as well as a thief) is in the candle'; this too is like the wolf devouring the sun or moon. Eclipses of sun or moon have been a terror to many heathen nations; the incipient and increasing obscuration of the luminous orb marks for them the moment when the gaping jaws of the wolf threaten to devour it, and they think by loud cries to bring it succour."[307] And again:—"The personality of the sun and moon shows itself moreover in a fiction that has well-nigh gone the round of the world. These two, in their unceasing unflagging career through the void of heaven, appear to be in flight, avoiding some pursuer. A pair of wolves are on their track, *Sköll* dogging the steps of the sun, *Hati* of the moon: they come of a giant race, the mightiest of whom, Mânagarmr (moon-dog), apparently but another name for Hati, is sure some day to *overtake and swallow the moon*."[308] Francis Osborn, whose *Advice* contains, in the opinion of Hallam, "a considerable sprinkling of sound sense and observation," thus counsels his son: "Imitate not the wild Irish or Welch, who, during eclipses, run about beating kettles and pans, thinking their clamour and vexations available to the assistance of the higher orbs."[309] "In eclipses of the moon, the Greenlanders carry boxes and kettles to the roofs of their houses, and beat on them as hard as they can."[310] With the Californian Indians, "on an eclipse, all is consternation. They congregate and sing, as some say to appease, and others to frighten, the evil spirits. They believe that the devils are eating up the luminary, and they do not cease until it comes forth in its wonted splendour."[311] Among certain Indian tribes "dogs were supposed to stand in some peculiar relation to the moon, probably because they howl at it, and run at night; uncanny practices which have cost them dear in reputation. The custom prevailed among tribes so widely asunder as Peruvians, Tupis, Creeks, Iroquois, Algonkins, and Greenland Eskimos, to thrash the curs most soundly during an eclipse. The Creeks explained this by saying that the big dog was swallowing the sun, and that by whipping the little ones they could make him desist. What the big dog was they were not prepared to say. We know. It was the night goddess, represented by the dog, who was thus shrouding the world at midday."[312]

It is well known that Columbus found his acquaintance with the calculations of astronomy of great practical value. For when, during his last expedition, he was reduced to famine by the inhabitants of the newly discovered continent, who kept him and his companions prisoners, he, aware that an eclipse was at hand, threatened to deprive them of the light of the moon, if they did not forthwith bring him provisions. At first they did not care; but when the moon disappeared, they brought abundance of supplies, with much entreaty of pardon. This occurred on the 1st day of March, 1504, a date which modern tables of lunar eclipses may fully verify.

"In the Mexican mythology we read of the woman serpent, or the moon, devoured by the sun, a myth probably descriptive of the changes in the phases of the moon."[313] More probably this myth referred to the moon's eclipse; for Bradford tells us that "the Mexicans believed when there was an eclipse of the sun or moon, that one of those bodies was being devoured by the other. The Peruvians believed these phenomena portended some great calamity; that the eclipsed body was sick and about to die, in which case the world would perish. As soon as an eclipse commenced, they made a dreadful noise with their musical instruments; they struck their dogs and made them howl, in the hope that the moon, which they believed had an affection for those animals in consequence of some signal service which they had rendered her, would have pity on their cries. The Araucanians called eclipses the 'deaths' of the sun and moon."[314] In Aglio we are told of the Mexicans that "in the year of Five Rabbits, or in 1510, there was an eclipse of the sun; they take no account of the eclipses of the moon, but only of those of the sun; for they say that the sun devours the moon when an eclipse of the moon takes place."[315] "The Tlascaltecs, regarding the sun and the moon as husband and wife, believed eclipses to be domestic quarrels. Ribas tells how the Sinaloas held that the moon in an eclipse was darkened with the dust of battle. Her enemy had come upon her, and a terrible fight, big with consequence to those on earth, went on in heaven. In wild excitement the people beat on the sides of their houses, encouraging the moon, and shooting flights of arrows up into the sky to distract her adversary. Much the same as this was also done by certain Californians."[316] "At a lunar eclipse the Orinoko Indians seized their hoes and laboured with exemplary vigour on their growing corn, saying the moon was veiling herself in anger at their habitual laziness."[317] The umbrated moon did good in this way: as many of us remember the beautiful comet of 1858 did good, when it frightened some trembling Londoners into a speedy settlement of old debts, in anticipation of the final account. Ellis says of the Tahitians: "An eclipse of the moon filled them with dismay; they supposed the planet was *natua*, or under the influence of the spell of some evil spirit that was destroying it. Hence they repaired to the temple, and offered prayers for the moon's release. Some imagined that on an eclipse, the sun and moon were swallowed by the god which they had by neglect offended. Liberal presents were offered, which were supposed to induce the god to abate his anger, and eject the luminaries of day and night from his stomach."[318] The Tongans or Friendly Islanders have a notion that the earth's surface is flat, that the sun and moon "pass through the sky and come back some way, they know not how. When the moon is eclipsed, they attribute the phenomenon to a thick cloud passing over it: the same with the sun."[319] In the Hervey Islands, the common exclamation during an eclipse is, "Alas! a divinity has devoured the moon!"

Finally, to close this chapter where it commenced, in Chaldaea, the cradle of *star-reading*, Sir Austen Henry Layard says: "I gained, as other travellers have done before me, some credit for wisdom and superhuman knowledge by predicting, through the aid of an almanack, a partial eclipse of the moon. It duly took place, to the great dismay of my guests, who well-nigh knocked out the bottoms of all my kitchen utensils in their endeavour to frighten away the jins who had thus laid hold of the planet. The common notion amongst ignorant Mahometans is, that an eclipse is caused by some evil spirit catching hold of the sun or moon. On such occasions, in Eastern towns, the whole population assembles with pots, pans, and other equally rude instruments of music, and, with the aid of their lungs, make a din and turmoil which might suffice to drive away a whole army of evil spirits, even at so great a distance."[320] We have reached three general conclusions. *First*, when the moon is occulted by the earth it is believed to be devoured by some evil demon, or by wolves or dogs. This is the superstitious vagary of the Hindoos, the Chinese, Asiatics generally, Europeans, Africans, Americans, and Polynesians. *Secondly*, a lunar eclipse is the precursor of some dreadful calamity to the inhabitants of the earth. This notion is also traceable in every quarter of the globe. And *thirdly*, during the obscuration the light of the moon is reddened, and at last extinguished, by the blood which flows from its wounds; which belief originates with the *Edda*, and obtains in the Western world. Students of sacred prophecy may still elect to deem these occurrences that are purely natural as of supernatural significance, and may risk the interests of true religion in their insane disregard of science; but the truth will remain, in spite of their misconceptions, that eclipses of the moon have no concern with the moral destiny of mankind.

Lunar Influences

The superficies of the earth being twice seven times that of the moon, what an influence the earth must exercise over its satellite! We may be unable to describe this influence in all of its effects; but we may observe its existence in some of its apparent signs. The moon not only turns while we turn, but its rotations on its axis keep exact time with its revolutions round our globe; it accompanies us as we encircle the sun, facing us all the while, never turning its back upon us; it waits on us like a link-bearer, or lackey; is our admiring Boswell, living and moving and having its being in the equability it derives from attending its illustrious master. An African sage once illustrated this philosophical principle of the greater controlling the less, by the following fine conundrum. "Why does the dog waggle his tail?" This problem, being beyond his auditors, was given up. The sage made answer, "Because the dog is bigger than the tail; else the tail would waggle the dog." It is alarming to contemplate the effect which the moon might have upon our august earth, if it were fourteen times larger instead of fourteen times smaller in extent of surface. As it is, Luna's influences are so many and so mighty, that we will require considerable space merely to set them in order, and to substantiate them with a few facts. We believe that most, if not all, of them, are the offspring of superstition; but we shall none the less find them in every land, in every age. In the nineteenth century as well as in the dark ages, in London as well as in the ends of the earth, men of all colours and clans are found turning their faces heavenward to read their duty and destiny in the oracular face of the moon. Many consult their almanacks more than their Bibles, and follow the lunar phases as their sole interpretation of the will of God.

Among those who worship the moon as a personal deity, whether beneficent or malign, its influences are of course welcomed or dreaded as the manifestations of supreme power. In South America, for example, "the Botocudos are said to give the highest rank among the heavenly bodies to Taru, the moon, as causing thunder and lightning and the failure of vegetables and fruits, and as even sometimes falling to the earth, whereby many men die."[321] So, in Africa, the emotions of the worshippers vary with their subjective views of their god. "Negro tribes seem almost universally to greet the new moon, whether in delight or disgust. The Guinea people fling themselves about with droll gestures, and pretend to throw firebrands at it; the Ashango men behold it with superstitious fear; the Fetu negroes jumped thrice into the air with hands together and gave thanks."[322] But even amongst men who neither personify nor deify the moon, its dominion over the air, earth, and sea, over human health and happiness, is held to be so all-important, that if the Maker and Monarch of all were jealous, as men count jealousy, such lunar fears and affections would be unpardonable sin.

Let us proceed to particulars, rising from inorganic nature to beings endowed with the highest instruments of life. Even the mineral kingdom is supposed to be swayed by the moon; for in Scotland, Martin says, "The natives told me, that the rock on the east side of Harries, in the Sound of Island Glass, hath a vacuity near the front, on the north-west side of the Sound; in which they say there is a stone that they call the *Lunar Stone*, which advances and retires according to the increase and decrease of the moon."[323] An ancient instance of belief in lunar influence upon inanimate matter is cited by Plutarch. "*Euthydemus* of *Sunium* feasted us upon a time at his house, and set before us a wilde bore, of such bignesse, that all wee at the table wondred thereat; but he told us that there was another brought unto him farre greater; mary naught it was, and corrupted in the carriage, by the beames of the moone-shine; whereof he made great doubt and question, how it should come to passe; for that he could not conceive, nor see any reason, but that the sunne should rather corrupt flesh, being as it was, farre hotter than the moone."[324] Pliny said that the moon corrupted carcases of animals exposed to its malefic rays. As with the lifeless, so with the living. "The inhabitants of St. Kilda observe that when the April moon goes far in May, the fowls are ten or twelve days later in laying their eggs than ordinarily they use to be."[325] The influence of the moon upon vegetation is an opinion hoary with age. In the *Zend-Avesta* we read, "And when the light of the moon waxes warmer, golden-hued plants grow on from the earth during the spring."[326] An old English author writes:—

> "Sowe peason and beanes, in the wane of the moone,
> Who soweth them sooner, he soweth too soone
> That they with the planet may rest and arise,
> And flourish, with bearing most plentiful wise."[327]

Cucumbers, radishes, turnips, leeks, lilies, horseradish, saffron, and other plants, are said to increase during the fulness of the moon; but onions, on the contrary, are much larger and are better nourished during the decline.[328] To recur to Plutarch is to find him saying: "The moone showeth her power most evidently even in those bodies, which have neither sense nor lively breath; for carpenters reject the timber of trees fallen in the ful-moone, as being soft and tender, subject also to the worme and putrifaction, and that quickly, by reason of excessive moisture; husbandmen, likewise, make haste to gather up their wheat and other grain from the threshing-floore, in the wane of the moone, and toward the end of the month, that being hardened thus with drinesse, the heape in the garner may keepe the better from being fustie, and continue the longer; whereas corne which is inned and laied up at the full of the moone, by reason of the softnesse and over-much moisture, of all other, doth most cracke and burst. It is commonly said also, that if a leaven be laied in the ful-moone, the paste will rise and take leaven better."[329] Still in Cornwall the people gather all their medicinal plants when the moon is of a certain age; which practice is very probably a relic of druidical superstition. "In some parts it is a prevalent belief that the growth of mushrooms is influenced by the changes of the moon, and in Essex the subjoined rule is often scrupulously adhered to:—

"When the moon is at the full,
Mushrooms you may freely pull
But when the moon is on the wane,
Wait ere you think to pluck again.'"[330]

Henderson says, "I may, perhaps, mention here, that apples are said to 'shrump up' in Devonshire if picked when the moon is waning."[331] A writer of miscellaneous literature tells us that "it has been demonstrated that moonlight has the power, *per se*, of awakening the sensitive plant, and consequently that it possesses an influence of some kind on vegetation. It is true that the influence is very feeble, compared with that of the sun; but the action is established, and the question remains, what is the practical value of the fact? 'It will immediately,' says Professor Lindley, 'occur to the reader that possibly the screens which are drawn down over hothouses at night, to prevent loss of heat by radiation, may produce some unappreciated injury by cutting off the rays of the moon, which nature intended to fall upon plants as much as the rays of the sun."[332] The same author says elsewhere, "Columella, Cato, Vitruvius, and Pliny, all had their notions of the advantages of cutting timber at certain ages of the moon; a piece of mummery which is still preserved in the royal ordonnances of France to the conservators of the forests, who are directed to fell oaks only 'in the wane of the moon' and 'when the wind is at north.'"[333] Of trees, astrologers affirm that the moon rules the palm tree (which the ancients say "sends forth a twig every time the moon rises") and all plants, trees, and herbs that are juicy and full of sap.[334]

"A description of the New Netherlands, written about 1650, remarks that the savages of that land 'ascribe great influence to the moon over crops.' This venerable superstition, common to all races, still lingers among our own farmers, many of whom continue to observe 'the signs of the moon' in sowing grain, setting out trees, cutting timber, and other rural avocations."[335] What is here said of the new world applies also to the old; for in England a current expression in Huntingdonshire is "a dark Christmas sends a fine harvest": dark meaning moonless.

Of the lunar influence upon the tides, old John Lilly writes: "There is nothing thought more admirable, or commendable in the sea, than the ebbing and flowing; and shall the moone, from whom the sea taketh this virtue, be accounted fickle for encreasing and decreasing?"[336] Another writer of the sixteenth century says, "The moone is founde, by plaine experience, to beare her greatest stroke uppon the seas, likewise in all things that are moiste, and by consequence in the braines of man."[337] Dennys tells us that "the influence exerted by the moon on tides is recognised by the Chinese."[338] What some record in prose, others repeat in rhyme. The following is *one* kind of poetry.

"Moone changed, keepes closet, three daies as a Queene,
Er she in hir prime, will of any be scene:
If great she appereth, it showreth out,
If small she appereth, it signifieth drout.
At change or at full, come it late, or else soone,
Maine sea is at highest, at midnight and noone,
But yet in the creekes, it is later high flood:
Through farnesse of running, by reason as good."[339]

Indirectly, through the influence upon the tides, the moon is concerned in human mortality.

"Tyde flowing is feared, for many a thing,
Great danger to such as be sick it doth bring.
Sea eb, by long ebbing, some respit doth give,
And sendeth good comfort, to such as shal live."[340]

Henderson says, "It is a common belief along the east coast of England, from Northumberland to Kent, that deaths mostly occur during the falling of the tide."[341] Every reader of the inimitable Dickens will be reminded here of the death of poor old Barkis.

"'He's a-going out with the tide,' said Mr. Peggotty to me,

behind his hand.

"My eyes were dim, and so were Mr. Peggotty's; but I repeated in a whisper, 'With the tide?'

"'People can't die, along the coast,' said Mr. Peggotty, 'except when the tide's pretty nigh out. They can't be born, unless it's pretty nigh in-not properly born, till flood. He's a-going out with the tide. It's ebb at half-arter three, slack water half an hour. If he lives till it turns, he'll hold his own till past the flood, and go out with the next tide.'

"'He's coming to himself,' said Peggotty.

"Mr. Peggotty touched me, and whispered with much awe and reverence, 'They are both a-going out fast.'

"He now opened his eyes.

"I was on the point of asking him if he knew me, when he tried to stretch out his arm, and said to me distinctly, with a pleasant smile,—

"'Barkis is willin'.'

"And, it being low water, he went out with the tide."[342]

That the rise and fall of our tides twice a day, with spring and neap tides twice in the lunar month, are the effect of the combined action of the sun and moon, is never called in question. The water under the moon is drawn up from the earth, and the earth is drawn from the water on the opposite side, the consequence of which is two high tides in the two hemispheres at the same hour. The rotation of the earth bringing the same point of the ocean twice under the moon's meridian, once under the upper meridian and once under the lower, each hemisphere has two high tides in the course of the day. The spring tide is caused by the attractive force of the sun and moon acting in conjunction, or in a straight line; and the neap tide is caused by the moon being in quadrature, or when the sun and moon are at right angles to each other. They counteract each other's influence, and our tides arc therefore low. So much is science; but the connection of ebb and flow with life and death is superstition.

From a very remote antiquity, in the twilight of natural

astrology, a belief arose that changes in the weather were occasioned by the moon.[343] That the notion lives on, and will not soon die, is clear to any one who is conversant with current literature and common folk-lore. Even intelligent, well-informed people lend it countenance. Professor Newcomb, of Washington, rightly says: "Thus far there is no evidence that the moon directly affects the earth or its inhabitants in any other way than by her attraction, which is so minute as to be entirely insensible except in the ways we have described. A striking illustration of the fallibility of the human judgment when not disciplined by scientific training is afforded by the opinions which have at various times obtained currency respecting a supposed influence of the moon on the weather. Neither in the reason of the case nor in observations do we find any real support for such a theory. It must, however, be admitted that opinions of this character are not confined to the uneducated."[344] Mr. Edward B. Tylor holds similar language: "The notion that the weather changes with the moon's quarterings is still held with great vigour in England. That educated people to whom exact weather records are accessible should still find satisfaction in the fanciful lunar rule, is an interesting case of intellectual survival."[345] No marvel that the "heathen Chinee" considers lunar observations as forecasting scarcity of provisions he is but of the same blood with his British brother, who takes his tea and sends him opium. "The Hakkas (and also many Puntis) believe that if in the night of the fifteenth day of the eighth month (mid autumn) there are clouds obscuring the moon before midnight, it is a sign that oil and salt will become very dear. If, however, there are clouds obscuring the moon after midnight, the price of rice will, it is supposed, undergo a similar change."[346]

One of our provincial proverbs is: "So many days old the moon is on Michaelmas Day, so many floods after." Sometimes a proverb is a short saying spoken after long experience; at other times it is a small crystal left after a lengthy evaporation. In certain instances our rural apothegms are sacred relics of extinct but canonized fictions. An equally wise prediction is that if Christmas comes during a waxing moon we shall have a very good year; and the nearer to the new moon, the better. But if during a waning moon, a hard year; and the nearer the end of the moon, so much the worse. Another sage belief is that the condition of the weather is dependent upon the day of the week upon which the new moon chances to fall. We are told that "Dr. Forster, of Bruges, well known as a meteorologist, declares that by the *Journal* kept by his grandfather, father, and self, ever since 1767, to the present time, whenever the new moon has fallen on a *Saturday*, the following *twenty days* have been wet and windy, in nineteen cases out of twenty."[347] In Italy it is said, "If the moon change on a Sunday, there will be a flood before the month is out." New moon on Monday, or moon-day, is, of course, everywhere held a sign of good weather and luck.

That a misty moon is a misfortune to the atmosphere is widely supposed. In Scotland it is an agricultural maxim among the canny farmers that—

> "If the moon shows like a silver shield,
> You need not be afraid to reap your field
> But if she rises haloed round,
> Soon we'll tread on deluged ground."[348]

Others say that a mist is unfavourable only with the new moon, not with the old.

> "An old moon in a mist
> Is worth gold in a kist (chest)
> But a new moon's mist
> Will never lack thirst,"[349]

is a rugged rhyme found in several places. In Cornwall the idea is that—

> "A fog and a small moon
> Bring an easterly wind soon."

The east wind, as we know, is dry. Two of the Shepherd of Banbury's rules are:

> "xii. If mists in the new moon, rain in the old.
> xiii. If mists in the old, rain in the new moon."[350]

One thing is a meteorological certainty: the full moon very frequently clears the sky. But this may be partly accounted for by the fact that a full moon shows the night to be clear, which in the moon's absence might be called cloudy.

Another observation shows that in proportion to the clearness of the night is its cold. The clouds covering the earth with no thick blanket, it radiates its heat into space. This has given rise to the notion that the moon itself reduces our temperature. It is *cold* at night without doubt. But the cold moon is so warm when the sun is shining full on its disk that no creature on earth could endure a moment's contact with its surface. The centre of the "pale-faced moon" is hotter than boiling water. This thought may cheer us when "the cold round moon shines deeply down." We may be pardoned if we take with a tincture of scepticism the following statement "Native Chinese records aver that on the 18th day of the 6th moon, 1590, snow fell one summer night from the midst of the moon. The flakes were like fine willow flowers on shreds of silk."[351] Instead of cold, it is more likely that the white moon gives us heat, for from Melloni's letter to Arago it seems to be already an ascertained fact. Having concentrated the lunar rays with a lens of over three feet diameter upon his thermoscopic pile, Melloni found that the needle had deviated from 0° 6' to 4° 8', according to the lunar phase. Other thermoscopes may give even larger indications; but meanwhile the Italian physicist has exploded an error with a spark of science.

"Another weather guide connected with the moon is, that to see 'the old moon in the arms of the new one' is reckoned a sign of fine weather; and so is the turning up of the horns of the new moon. In this position it is supposed to retain the water, which is imagined to be in it, and which would run out if the horns were turned down."[352] On this novel idea of a lunar bason or saucer, Southey writes from "Keswick, December 29th, 1828," as follows:— "Poor Littledale has this day explained the cause of our late rains, which have prevailed for the last six weeks, by a theory which will probably be as new to you as it is to me. 'I have observed,' he says, 'that, when the moon is turned upward, we have fine weather after it; but if it is turned down, then we have a wet season; and the reason I think is, that when it is turned down, it holds no water, like a bason, you know, and then down it all comes.' There, it will be a long while before the march of intellect shall produce a theory as original as this, which I find, upon inquiry, to be the popular opinion here."[353] George Eliot has taken notice of this fancy in the burial of "poor old Thias Bede." "They'll ha' putten Thias Bede i' the ground afore ye get to the churchyard," said old Martin, as his son came up. "It 'ud ha' been better luck if they'd ha' buried him i' the forenoon when the rain was fallin'; there's no likelihoods of a drop now, an' the moon lies like a boat there, dost see? That's a sure sign o' fair weather; there's a many as is false, but that's sure."[354]

In Dekker's *Match Me in London*, Act i., the King says, "My Lord, doe you see this change in the moone? Sharp hornes doe threaten windy weather."

In the famous ballad of Sir Patrick Spens, concerning whose origin there has been so much discussion, without eliciting any very accurate information, we read:

"O ever alack! my master dear,
  I fear a deadly storm.
I saw the new moon late yestreen,
  Wi' the auld moon in her arm
And if ye gang to sea, maister,
  I fear we'll suffer harm."[355]

Jamieson informs us that "prognostications concerning the weather, during the course of the month, are generally formed by the country people in Scotland from the appearance of the *new moon*. It is considered as an almost infallible presage of bad weather, if she *lies sair on her back*, or when her horns are pointed towards the zenith. It is a similar prognostic, when the new moon appears *wi' the auld moon in her arm*, or, in other words, when that part of the moon which is covered with the shadow of the earth is seen through it."[356] The last sentence is a *lapsus calami*. Dr. Jamieson should have said, when that part of the moon which is turned from the sun is dimly visible through the reflected light of the earth.

"At Whitby, when the moon is surrounded by a halo with watery clouds, the seamen say that there will be a change of weather, for the 'moon dogs' are about."[357] At Ulceby, in Lincolnshire, "there is a very prevalent belief amongst sailors and seafaring men that when a large star or planet is seen near the moon, or, as they express it, 'a big star is dogging the moon,' that this is a certain prognostication of wild weather. I have met old sailors having the strongest faith in this prediction, and who have told me that they have verified it by a long course of observation."[358]

"Some years ago," says a writer from Torquay, "an old fisherman of this place told me, on the morning next after a violent gale, that he had foreseen the storm for some time, as he had observed one star ahead of the moon, towing her, and another astern, chasing her. 'I know'd 'twas coming, safe enough.'"[359] The moon was simply in apparent proximity to two stars; but the old Devonian descried mischief.

The following incident from Zulu life will be of interest. "1878. A curious phenomenon occurred 7th January. A bright star appeared near the moon at noonday, the sun shining brightly. *Omen*—The natives from this foretold the coming war with the Amazulu. Intense heat and drought prevailed at this time."[360]

Hitherto we have reviewed only the imaginary influences of the moon over inanimate nature and what are called irrational beings. We have seen that this potent orb is supposed to affect the lightning and thunder of the air; the rocks and seas, the vegetables and animals of the earth; and generally to govern terrestrial matters in a manner altogether its own. Furthermore, we have found these imaginations rooted in all lands, and among men whose culture might have been expected to refuse such fruitless excrescences. When classical authors counsel us to set eggs under the hen at new moon, and to root up trees only when the moon is waning and after mid-day; and when "the wisest, brightest," if not the "meanest of mankind" seriously attributes to the moon the extraction of heat, the furtherance of putrification, the increase of moisture, and the excitement of animal spirits, with the increase of hedges and herbs if cut or set during certain phases of that body, we can but repeat to ourselves the saying, "The best of men are but men at the best." The half, however, has not been told; and we must now pass on to speak of lunar influences upon the birth, health, intellect, and fortune of microcosmical man.

In the system of astrology, which professed to interpret the events of human existence by the movements of the stars, the moon was one of the primary planets. As man was looked upon in the light of a microcosm, or world in miniature, so the several parts of his constitution were viewed as but a reproduction in brief of the great parts of the vast organism. Creation was a living, intelligent being, whose two eyes were the sun and the moon, whose body was the earth, whose intellect was the ether, whose wings were the heavens. Man was an epitome of all this; and as the functions of the less were held to correspond with the functions of the greater, the microcosm with the macrocosm, man's movements could be inferred by first ascertaining the motions of the universe. The moon, having dominion in the twelve "houses" of heaven, through which she passed in the course of the year, her *aspects* to the other bodies were considered as of prime significance, in indicating benignant or malignant influences upon human life. This system, which was based upon ignorance and superstition, and upheld by arbitrary rules and unreasoning credulity, is so repugnant to all principles of science and common sense, that it would be unworthy of notice, if we did not know that to this day there are educated persons still to be seen poring over old almanacs and peering into the darkness of divination, to read their own fortune or that of their children by the dim light of some lucky or unlucky configuration of the planets with the moon. The wheel of fortune yet revolves, and the despotism of astrology is not dead. The lunar influence is considered supreme in the hour of birth. Nay, with some the moon is potential even before birth. In Iceland it is said: "If a pregnant woman sit with her face turned towards the moon, her child will be a lunatic."[361] And this imagination obtains at home as well as abroad. We are told that "astrologers ascribe the most powerful influence to the moon on every person, both for success and health, according to her zodiacal and mundane position at birth, and her aspects to other planets. The sensual faculties depend almost entirely on the moon, and as she is aspected so are the moral or immoral tendencies. She has great influence always upon every person's constitution."[362] This is the doctrine of a book published not thirty years ago. Another work, issued also in London, says, "Cynthia, 'the queen of heaven,' as the ancients termed her, or the MOON, the companion of the earth, and chief source of our evening light, is a cold, moist, watery, phlegmatic planet, variable to an extreme, in astrological science; and partaking of good or evil, as she is aspected by good or evil stars. When angular and unafflicted in a nativity, she is the promissory pledge of great success in life and continual good fortune. She produces a full stature, fair, pale complexion, round face, gray eyes, short arms, thick hands and feet, smooth, corpulent, and phlegmatic body. Blemishes in the eyes, or a peculiar weakness in the sight, is the result of her being afflicted by the Sun. Her conjunction, semi-sextile, sextile, or trine, to Jupiter, is exceeding fortunate; and she is said by the old Astrologers to govern the *brain*, *stomach*, *bowels*, *left eye* of the male, and *right eye* of the female. Her usual diseases are rheumatism, consumption, palsy, cholic, apoplexy, vertigo, lunacy, scrophula, smallpox, dropsy, etc.; also most diseases peculiar to young children."[363] Such teaching is not a whit in advance of Plutarch's odd dictum that the moon has a "special hand in the birth of children."

If this belief have disciples in London, it is not by any means confined to that city. In Sweden great influence is ascribed to the moon, not only in regulating the weather, but as affecting all the affairs of man's daily life. The lower orders, and many of the better sort, will not fell a tree for agricultural purposes in the wane of that orb, lest it should shrink and decay; nor will the housewife then slaughter for her family, lest the meat should shrivel and melt away in the pot. The moon is the domestic deity, whom the household must fear: the Fortuna who presides over the daily doings of sublunary mortals. In the matter of birth, we find Francis Bacon affirming that "the calculation of nativities, fortunes, good or bad hours of business, and the like fatalities, are mere levities that have little in them of certainty and solidity, and may be plainly confuted by physical reasons";[364] and yet in his Natural History he writes: "It may be that children and young cattle that are brought forth in the full of the moon, are stronger and larger than those that are brought forth in the wane."[365] There surely can be no superstition in studying the moon's conjunctions and oppositions if her influence in a nativity have the slightest weight. And this influence is still widely maintained by philosophers who read Bacon, as well as by the peasants who read nothing at all. "In Cornwall, when a child is born in the interval between an old moon and the first appearance of a new one, it is said that it will never live to reach the age of puberty. Hence the saying, 'no moon, no man.' In the same county, too, when a boy is born in the wane of the moon, it is believed that the next birth will be a girl, and vice versa; and it is also commonly said that when a birth takes place on the 'growing of the moon' the next child will be of the same sex."[366]

As a natural proceeding, we find that the moon has influence when the child is weaned. Caledonian mothers very carefully observe the lunar phases on this account. Jamieson tells us that "this superstition, with respect to the fatal influence of a waning moon, seems to have been general in Scotland. In Angus, it is believed, that, if a child be put from the breast during the waning of the moon, it will decay all the time that the moon continues to wane."[367] So in the heart of Europe, "the Lithuanian precept to wean boys at a waxing, but girls on a waning moon, no doubt to make the boys sturdy and the girls slim and delicate, is a fair match for the Orkney Islanders' objection to marrying except with a growing moon, while some even wish for a flowing tide."[368] As to marriage, the ancient Greeks considered the day of the full moon the most propitious period for that ceremony. In Euripides, Clytemnestra having asked Agamemnon when he intended to give Iphigenia in marriage to Achilles, he replies, "When the full moon comes forth with good luck." In Pindar, too, this season is preferred.[369]

Lunar influences over physical health and disease must be a fearful contemplation to those who are of a superstitious turn. There is no malady within the whole realm of pathology which the moon's destroying angel cannot inflict; and from the crown of the head to the sole of the foot the entire man is at the mercy of her beams. We have all seen those disgusting woodcuts to which the following just condemnation refers: "The moon's influence on parts of the human body, as given in some old-fashioned almanacs, is an entire *fallacy*; it is most untrue and absurd, often indecent, and is a discredit to the age we live in."[370] Most of these inartistic productions are framed upon the assumption of the old alchymists that the physiological functions were regulated by planetary influence. The sun controlled the heart, the moon the brain, Jupiter the lungs, Saturn the spleen, Mars the liver, Venus the kidneys, and Mercury the reproductive powers. But even with this distribution among the heavenly bodies the moon was allowed plenipotentiary sway. As in mythology it is the god or goddess of water, so in astrology it is the embodiment of moisture, and therefore rules the humours which circulate throughout the human system. No wonder that phlebotomy prevailed so long as the reign of the moon endured. "This lunar planet," says La Martinière, "is damp of itself, but, by the radiation of the sun, is of various temperaments, as follows: in its first quadrant it is warm and damp, at which time it is good to let the blood of sanguine persons; in its second it is warm and dry, at which time it is good to bleed the choleric; in its third quadrant it is cold and moist, and phlegmatic people may be bled; and in its fourth it is cold and dry, at which time it is well to bleed the melancholic." Whatever the moon's phase may be, let blood be shed! We are reminded here of that sanguifluous theology, which even Christians of a certain temperament seem to enjoy, while they sing of fountains filled with blood: as though a God of love could take delight in the effusion of precious life. La Martinière continues, and physicians will make a note of his words: "It is a thing quite necessary to those who meddle with medicine to understand the movement of this planet, in order to discern the causes of sickness. And as the moon is often in conjunction with Saturn, many attribute to it apoplexy, paralysis, epilepsy, jaundice, hydropsy, lethargy, catapory, catalepsy, colds, convulsions, trembling of the limbs, etc., etc. I have noticed that this planet has such enormous power over living creatures, that children born at the first quarter of the declining moon are more subject to illness, so that children born when there is no moon, if they live, are weak, delicate, and sickly, or are of little mind or idiots. Those who are born under the house of the moon which is Cancer, are of a phlegmatic disposition."[371]

That the ancient Hebrews, Greeks, and Romans believed in the deleterious influence of the moon on the health of man, is very evident. The Talmud refers the words, "Though I walk through the valley of the shadow of death" (Ps. xxiii. 4) "to him who sleeps in the shadow of the moon."[372] Another Psalm (cxxi. 6) reads, literally, "By day the sun shall not smite thee, and the moon in the night." In the Greek Testament we find further proof of this belief. Among those who thronged the Great Teacher (Matt. iv. 24) were the σεληνιαζομένοι (*lunatici*, Beza; *i lunatici*, Diodati; *les lunatiques*, French version; "those who were lunatick"). The Revised Version of 1881 reads "epileptic," but that is a comment, not a translation. So again (Matt. xvii. 15) we read of a boy who was "lunatick"—σεληνιάζεται. On which Archbishop Trench remarks, "Of course the word originally, like μανία (from μηνη) and lunaticus, arose from the widespread belief of the evil influence of the moon on the human frame."[373] Jerome attributes all this superstition to daemons, of which men were the dupes. "The *lunatics*," he says, "were not really smitten by the moon, but were believed to be so, through the subtlety of the daemons, who by observing the seasons of the moon sought to bring an evil report against the creature, that it might redound to the blasphemy of the Creator."[374] Demons or no demons, faith in moonstroke is clear enough. Pliny was of opinion that the moon induced drowsiness and stupor in those who slept under her beams. Galen, in the second century, taught that those who were born when the moon was falciform, or sickle-shaped, were weak and short-lived, while those born during the full moon were vigorous and of long life. He also took notice of the lunar influence in epilepsy[375] of which fearful malady a modern physician writes, "This disease has been known from the earliest antiquity, and is remarkable as being that malady which, even beyond insanity, was made the foundation of the doctrine of possession by evil spirits, alike in the Jewish, Grecian, and Roman philosophy."[376] The terrible disorder was a fact; and evil spirits or the moon had to bear the blame.

In modern times the moon is no less the deity of insalutary disaster. Of Mexico, Brinton says: "Very different is another aspect of the moon-goddess, and well might the Mexicans paint her with two colours. The beneficent dispenser of harvests and offspring, she nevertheless has a portentous and terrific phase. She is also the goddess of the night, the dampness, and the cold; she engenders the miasmatic poisons that rack our bones; she conceals in her mantle the foe who takes us unawares; she rules those vague shapes which fright us in the dim light; the causeless sounds of night or its more oppressive silence are familiar to her; she it is who sends dreams wherein gods and devils have their sport with man, and slumber, the twin brother of the grave."[377] So farther south, "the Brazilian mother carefully shielded her infant from the lunar rays, believing that they would produce sickness; the hunting tribes of our own country will not sleep in its light, nor leave their game exposed to its action. We ourselves have not outgrown such words as lunatic, moon-struck, and the like. Where did we get these ideas? The philosophical historian of medicine, Kurt Sprengel, traces them to the primitive and popular medical theories of ancient Egypt, in accordance with which all maladies were the effects of the anger of the goddess Isis, the moisture, the moon."[378] Perhaps Dr. Brinton's own Mexican myth is a better elucidation of this origin of nocturnal evil than that which traces it to Egypt. According to an ancient tradition in Mexico, "it is said that in the absence of the sun all mankind lingered in darkness. Nothing but a human sacrifice could hasten his arrival. Then Metzli, the moon, led forth one Nanahuatl, the leprous, and building a pyre, the victim threw himself in its midst. Straightway Metzli followed his example, and as she disappeared in the bright flames, the sun rose over the horizon. Is not this a reference to the kindling rays of the aurora, in which the dark and baleful night is sacrificed, and in whose light the moon presently fades away, and the sun comes forth?"[379] We venture to think that it is, and that it is nearest to a natural explanation of purely natural effects.

Coming next to Britain, we find that "no prejudice has been more firmly rivetted than the influence of the moon over the human frame, originating perhaps in some superstition more ancient than recorded by the earliest history. The frequent intercourse of Scotland with the north may have conspired to disseminate or renew the veneration of a luminary so highly venerated there, in counteracting the more southern ecclesiastical ordinances."[380] Forbes Leslie surely goes too far, and mixes matters up too much, when he writes: "An ancient belief, adhered to by the ignorant after being denounced and apparently disproved by the learned, is now admitted to be a fact; viz. the influence of the moon in certain diseases. This, from various circumstances, is more apparent in some of the Asiatic countries, and may have given rise to the custom which extended into Britain, of exposing sick children on the housetops."[381] We know that the *solar* rays, from the time of

Hippocrates, the reputed "father of medicine," were believed by the Greeks to prolong life; and that the Romans built terraces on the tops of their houses called *solaria*, where they enjoyed their solar baths. "Levato sole levatur morbus," was one of their medical axioms. But who ever heard of the *lunar* rays as beneficial? If sick children were exposed on the housetops, it must have been in the daytime; and, unless it were intended as an alterative, it is difficult to see what connection this had with the belief that disease was the product of the lunar beam. Besides, is the moon's influence in disease an admitted fact? The "certain diseases" should be specified, and their lunar origin sustained.

The following strange superstition is singularly like that interpolated legend in the Gospel of John, about the angel troubling the pool of Bethesda. In this case the medicinal virtue seems to come with the change of the moon. But in both cases supernatural agency is equally mythical. "A cave in the neighbourhood of Dunskey ought also to be mentioned, on account of the great veneration in which it is held by the people. At the change of the moon (which is still considered with superstitious reverence), it is usual to bring, even from a great distance, infirm persons, and particularly ricketty children, whom they often suppose bewitched, to bathe in a stream which pours from the hill, and then dry them in the cave."[382]

Those who are in danger of apoplexy, or other cerebral disease, through indulgence too freely in various liquids, vinous and spirituous, should cherish Bacon's sapient deliverance: "It is like that the brain of man waxeth moister and fuller upon the full of the moon; and therefore it were good for those that have moist brains, and are great drinkers, to take sume of *lignum aloes*, rosemary, frankincense, etc., about the full of the moon. It is like, also, that the humours in men's bodies increase and decrease as the moon doth; and therefore it were good to purge some day or two after the full; for that then the humours will not replenish so soon again."[383] All this sounds so unphilosophical that it is almost incredible that the learned Bacon believed what he wrote. Darker superstitions, however, still linger in our land. "In Staffordshire, it is commonly said, if you want to cure chin-cough, take out the child and let it look at the new moon; lift up its clothes and rub your right hand up and down its stomach, and repeat the following lines (looking steadfastly at the moon, and rubbing at the same time):—

'What I see, may it increase;
What I feel, may it decrease;

In the name of the Father, Son, and Holy Ghost. Amen.'"[384] There is a little ambiguity here. What is felt is the child's stomach. But the desire is not that that may decrease, but only the whooping cough, which is *felt*, we take it, by proxy. A lady, writing of the southern county of Sussex, says: "A superstition lingering amongst us, worthy of the days of paganism, is that the new May moon, aided by certain charms, has the power of curing scrofulous complaints."[385]

As the cutting of hair, finger-nails, and corns has some relation to health and comfort, we may here mention that in Devonshire it is said that hair and nails should always be cut in the waning of the moon, thereby beneficial consequences will result. If corns are cut after the full moon, some say that they will gradually disappear. In the *British Apollo* we have the following request for advice:

"Pray tell your querist if he may
Rely on what the vulgar say,
That when the moon's in her increase,
If corns be cut they'll grow apace
But if you always do take care
After the full your corns to pare,
They do insensibly decay
And will in time wear quite away.
If this be true, pray let me know,
And give the reason why 'tis so."[386]

The following passage is worth quoting, without any abbreviation, as an excellent summary of wisdom and sense regarding the moon's influence on health: "There is much reason for regarding the moon as a source of evil, yet not that she herself is so, but only the circumstances which attend her. With us it happens that a bright moonlight night is always a cold one. The absence of cloud allows the earth to radiate its heat into space, and the air gradually cools, until the moisture it contained is precipitated in the form of dew, and lies like a thick blanket on the ground to prevent a further cooling. When the quantity of moisture in the air is small, the refrigerating process continues until frost is produced, and many a moonlight night in spring destroys half or even the whole of the fruit of a new season. Moonlight, therefore, frequently involves the idea of frigidity. With us, whose climate is comparatively cold, the change from the burning, blasting, or blighting heat of day, or sun-up, to the cold of a clear night, or sun-down, is not very great, but within the tropics the change is enormous. To such sudden vicissitudes in temperature, an Indian doctor, in whom I have great confidence, attributes fevers and agues. As it is clear that those persons only, whose business or pleasure obliges them to be out on cloudless nights, suffer from the severe cold produced by the rapid radiation into space of the heat of their own bodies and that of the earth, those who remain at home are not likely to suffer from the effects of the sudden and continued chill. Still further, it is clear that people in general will not care to go out during the darkness of a moonless night, unless obliged to do so. Consequently few persons have experience of the deleterious influence of starlight nights. But when a bright moon and a hot, close house induce the people to turn out and enjoy the coldness and clearness of night, it is very probable that refrigeration may be followed by severe bodily disease. Amongst such a people, the moon would rather be anathematised than adored. One may enjoy half an hour, or perhaps an hour, of moonlight, and yet be blighted or otherwise injured by a whole night of it."[387] In Denmark a superstition is current concerning the noxious influences of night.

The Danes have a kind of elves which they call the "Moon Folk." "The man is like an old man with a low-crowned hat upon his head; the woman is very beautiful in front, but behind she is hollow, like a dough-trough, and she has a sort of harp on which she plays, and lures young men with it, and then kills them. The man is also an evil being, for if any one comes near him he opens his mouth and breathes upon them, and his breath causes sickness. It is easy to see what this tradition means: it is the damp marsh wind, laden with foul and dangerous odours; and the woman's harp is the wind playing across the marsh rushes at nightfall."[388] It is the Queen of the Fairies in the *Midsummer-Night's Dream* who says to the Fairy King,—

> These are the forgeries of jealousy
> And never, since the middle summer's spring,
> Met we on hill, in dale, forest or mead,
> By pavèd fountain, or by rushy brook,
> Or in the beachèd margent of the sea,
> To dance our ringlets to the whistling wind,
> But with thy brawls thou hast disturb'd our sport.
> No night is now with hymn or carol blest:
> Therefore the moon, the governess of floods,
> Pale in her anger, washes all the air,
> That rheumatic diseases do abound
> And this same progeny of evils comes
> From our debate, from our dissension
> We are their parents and original.

It will be thought rashly iconoclastic if we cast the least doubt upon the idea that blindness is caused directly by the light of the moon. So many cases have been adduced that it is considered a settled point. We, however, dare to dispute some of the evidence. For instance "A poor man born in the village *Rowdil*, commonly called St. Clement's, blind, lost his sight at every change of the moon, which obliged him to keep his bed for a day or two, and then he recovered his sight."[389] If logic would enable us to prove a negative to this statement, we would meet it with simple denial. But we have no hesitation in saying that an investigation into this case would have exonerated the moon of any share in the affliction, and have revealed some other and likely cause. Our chief objection to this story is its element of periodicity; and we would require overwhelming testimony to establish even the probability of such a miracle once a month. That permanent injury may accrue to those whose sleeping eyes are exposed all night to the brightness of a full moon is probable enough. But this would take place not because the moon's beams were peculiarly baneful, but because any strong light would have a hurtful effect upon the eyes when fixed for hours in the condition of sleep.

We can quite believe that in a dry atmosphere like that of Egypt, where ophthalmia is very prevalent on account of constant irritation from the fine sand in the air, the eye, weary with the heat and aridity of the day, would be impaired if uncovered in the air to the rays of the moon. Carne's statements are consequently quite credible. He tells us: "The effect of the moonlight on the eyes in this country is singularly injurious; the natives tell you, as I found they also afterwards did in Arabia, always to cover your eyes when you sleep in the open air. The moon here really strikes and affects the sight, when you sleep exposed to it, much more than the sun; indeed, the sight of a person who should sleep with his face exposed at night, would soon be utterly impaired or destroyed."[390] For the same reason, that strong light oppresses the slumbering eye, "the seaman in his hammock takes care not to face the full moon, lest he be struck with blindness."[391] Nor can we regard the following as "an *extraordinary* effect of moonlight upon the human subject." In 1863, "a boy, thirteen years of age, residing near Peckham Rye, was expelled his home by his mother for disobedience. He ran away to a cornfield close by, and, on lying down in the open air, fell asleep. He slept throughout the night, which was a moonlight one. Some labourers on their way to work, next morning, seeing the boy apparently asleep, aroused him; the lad opened his eyes, but declared he could not see. He was conveyed home, and medical advice was obtained; the surgeon affirmed that the total loss of sight resulted from sleeping in the moonlight."[392] This was sad enough; but it was antecedently probable. No doubt a boy of thirteen who for disobedience was cast out of home in such a place as London had a hard lot, and went supperless to his open bed. His optic nerves were young and sensitive, and the protracted light so paralysed them that the morning found them closed "in endless night." This was a purely natural result: to admitting it, reason opposes no demur. But we must object, for truth's sake, to the tendency to account for natural consequences by assigning supernatural causes. The moon is no divinity; moonlight is no Divine emanation, with a vindictive animus; and those who countenance such silly superstition as that moonstroke is a mysterious, evil agency, are contributing to a polytheism which leads to atheism: for many gods logically means no GOD at all.

Another branch of this umbrageous if not fructuous tree of lunar superstition is the moon's influence on human fortune. Butler satirizes the visionary who—

"With the moon was more familiar
Than e'er was almanac well-willer (compiler);
Her secrets understood so clear
That some believed he had been there;
Knew when she was in fittest mood
For cutting corns, or letting blood:
Whether the wane be, or increase,
Best to set garlick, or sow pease:
Who first found out the man i' th' moon,
That to the ancients was unknown."—*Hudibras.*

A Swiss theologian amusingly describes the superstitious person who reads his fortune in the stars. He, it is said, "will be more afraid of the constellation fires than the flames of his next neighbour's house. He will not open a vein till he has asked leave of the planets. He will not commit his seed to the earth when the soil, but when the moon, requires it. He will have his hair cut when the moon is either in *Leo*, that his locks may stare like the lion's shag, or in *Aries*, that they may curl like a ram's horn. Whatever he would have to grow, he sets about when she is in her increase; but for what he would have made less, he chuses her wane. When the moon is in *Taurus*, he never can be persuaded to take physic, lest that animal which chews its cud should make him cast it up again. He will avoid the sea whenever *Mars* is in the midst of heaven, lest that warrior-god should stir up pirates against him. In *Taurus* he will plant his trees, that this sign, which the astrologers are pleased to call *fixed*, may fasten them deep in the earth. If at any time he has a mind to be admitted into the presence of a prince, he will wait till the moon is in conjunction with the sun; for 'tis then the society of an inferior with a superior is salutary and successful."[393]

The *new moon* is considered pre-eminently auspicious for commencements,—for all kinds of building up, and beginning *de novo*. Houses are to be erected and moved into; marriages are to be concluded, money counted, hair and nails cut, healing herbs and pure dew gathered, all at the new moon. Money counted at that period will be increased. The *full moon* is the time for pulling down, and thinking of the end of all things. Cut your timber, mow your grass, make your hay, not while the sun shines, but while the moon wanes; also stuff your feather-bed then, and so kill the newly plucked feathers completely, and bring them to rest. Wash your linen, too, by the waning moon, that the dirt may disappear with the dwindling light.[394] According to one old notion it was deemed unlucky to assume a new dress when the moon was in her decline. So says the Earl of Northampton: "They forbidde us when the moone is in a fixed signe, to put on a newe garment. Why so? Because it is lyke that it wyll be too longe in wearing, a small fault about this towne, where garments seldome last till they be payd for. But thyr meaning is, that the garment shall continue long, not in respect of any strength or goodness in the stuffe, but by the durance or disease of him that hath neyther leysure nor liberty to weare it."[395] It is well known that the ancient Hebrews held the new moon in religious reverence. The trumpets were blown, solemn sacrifices were offered and festivals held; and the first clay of the lunar month was always holy. In a Talmudic compilation, to which Dr. Farrar has contributed a preface, we find an interesting account of the *Blessing the new moon*. "It is a very pious act to bless the moon at the close of the Sabbath, when one is dressed in his best attire and perfumed. If the blessing is to be performed on the evening of an ordinary week-day, the best dress is to bc worn. According to the Kabbalists the blessings upon the moon are not to be said till seven full days after her birth, but, according to later authorities, this may be done after three days. The reason for not performing this monthly service under a roof, but in the open air, is because it is considered as the reception of the presence of the Shekinah, and it would not be respectful so to do anywhere but in the open air. It depends very much upon circumstances when and where the new moon is to be consecrated, and also upon one's own predisposition, for authorities differ. We will close these remarks with the conclusion of the Kitzur Sh'lu on the subject, which, at p. 72, col. 2, runs thus:

"When about to sanctify the new moon, one should straighten his feet (as at the Shemonah-esreh) and give one glance at the moon before he begins to repeat the ritual blessing, and having commenced it he should not look at her at all. Thus should he begin—'In the united name of the Holy and Blessed One' and His Shekinah, through that Hidden and Consecrated One! and in the name of all Israel!' Then he is to proceed with the 'Form of Prayer for the New Moon,' word for word, with out haste, but with solemn deliberation, and when he repeats—

'Blessed is thy Former, Blessed is thy Maker,
Blessed is thy Possessor, Blessed is thy Creator,'

he is to meditate on the initials of the four Divine epithets, which form 'Jacob'; for the moon, which is called 'the lesser light,' is his emblem or symbol, and he is also called 'little' (see Amos vii. 2). This he is to repeat three times. He is to skip three times while repeating thrice the following sentence, and after repeating three times forwards and backwards: thus (*forwards*)—'Fear and dread shall fall upon them by the greatness of thine arm; they shall be as still as a stone'; thus (*backwards*)—'Still as a stone may they be; by the greatness of thine arm may fear and dread fall on them'; he then is to say to his neighbour three times, 'Peace be unto you,' and the neighbour is to respond three times, 'Unto you be peace.' Then he is to say three times (very loudly), 'David, the King of Israel, liveth and existeth!' and finally, he is to say three times, 'May a good omen and good luck be upon us and upon all Israel! Amen!'"[396a]

That the ancient Germans held the moon in similar regard we know from Caesar, who, having inquired why Ariovistus did not come to an engagement, discovered this to be the reason: "that among the Germans it was the custom for their matrons to pronounce from lots and divination, whether it were expedient that the battle should be engaged in or not; that they had said, 'that it was not the will of heaven that the Germans should conquer, if they engaged in battle before the new moon.'"[396b]

Halliwell has reproduced an illustration of British superstition of the same sort. "A very singular divination practised at the period of the harvest moon is thus described in an old chap-book. When you go to bed, place under your pillow a prayer-book open at the part of the matrimonial service 'with this ring I thee wed'; place on it a key, a ring, a flower, and a sprig of willow, a small heart-cake, a crust of bread, and the following cards:—the ten of clubs, nine of hearts, ace of spades, and the ace of diamonds. Wrap all these in a thin handkerchief of gauze or muslin, and on getting into bed, cross your hands, and say:—

'Luna, every woman's friend,
To me thy goodness condescend
Let me this night in vision see
Emblems of my destiny.'

If you dream of storms, trouble will betide you; if the storm ends in a fine calm, so will your fate; if of a ring or the ace of diamonds, marriage; bread, an industrious life; cake, a prosperous life; flowers, joy; willow, treachery in love; spades, death; diamonds, money; clubs, a foreign land; hearts, illegitimate children; keys, that you will rise to great trust and power, and never know want; birds, that you will have many children; and geese, that you will marry more than once."[397] Such ridiculous absurdities would be rejected as apocryphal if young ladies were not still in the habit of placing bits of wedding cake under their pillows in the hope that their dreaming eyes may be enchanted with blissful visions of their future lords.

Hone tells us that in Berkshire, "at the first appearance of a new moon, maidens go into the fields, and, while they look at it, say:—

'New moon, new moon, I hail thee!
By all the virtue in thy body.
Grant this night that I may see
He who my true love is to be.'

Then they return home, firmly believing that before morning their future husbands will appear to them in their dreams."[398]

In Devonshire also "it is customary for young people, as soon as they see the first new moon after midsummer, to go to a stile, turn their back to it, and say:—

'All hail, new moon, all hail to thee!
I prithe, good moon, reveal to me
This night who shall my true love be
Who is he, and what he wears,
And what he does all months and years.'"[399]

Aubrey says the same of the Scotch of his day, and the custom is not yet extinct. "In Scotland (especially among the Highlanders) the women doe make a curtsey to the new moon; I have known one in England doe it, and our English woemen in the country doe retain (some of them) a touch of this gentilisme still, *e.g.*:—

'All haile to thee, moon, all haile to thee
I prithe, good moon, declare to me,
This night, who my husband must be.'

This they doe sitting astride on a gate or stile the first evening the new moon appears. In Herefordshire, etc., the vulgar people at the prime of the moon say, ''Tis a fine moon, God bless her.'"[400] "In Ireland, at the new moon, it is not an uncommon practice for people to point with a knife, and after invoking the Holy Trinity, to say:—

'New moon, true morrow, be true now to me,
That I ere the morrow my true love may see.'

The knife is then placed under the pillow, and silence strictly observed, lest the charm should be broken."[401]

Dr. Charles Mackay quotes from Mother Bridget's *Dream and Omen Book* the following prescription for ascertaining the events of futurity. "*First new moon of the year.* On the first new moon in the year take a pint of clear spring water, and infuse into it the *white* of an egg laid by a *white* hen, a glass of *white* wine, three almonds peeled *white*, and a tablespoonful of *white* rose-water. Drink this on going to bed, not making more nor less than three draughts of it; repeating the following verses three several times in a clear distinct voice, but not so loud as to be overheard by anybody:—

'If I dream of water pure
  Before the coming morn,
'Tis a sign I shall be poor,
  And unto wealth not born.

If I dream of tasting beer,
Middling, then, will be my cheer—
Chequered with the good and bad,
Sometimes joyful, sometimes sad;
But should I dream of drinking wine,
Wealth and pleasure will be mine.
The stronger the drink, the better the cheer—
Dreams of my destiny, appear, appear!'"[402]

The day of the week on which the moon is new or full, is a question that awakens the most anxious concern. In the north of Italy Wednesday is dreaded for a lunar change, and in the south of France the inauspicious day is Friday.[403] In most of our own rural districts Friday's new moon is much disliked

"Friday's moon,
Come when it wool,
It comes too soon."

Saturday is unlucky for the *new*, and Sunday for the *full* moon. In Norfolk it is said:—

"Saturday's new and Sunday's full,
Never was good, and never wull."

An apparently older version of the same weather-saw runs:—

"A Saturday's change, and a Sunday's prime,
Was nivver a good mune in nea man's time."

In Worcestershire, a cottager near Berrow Hill told Mr. Edwin Lees, F.L.S., that as the new moon had fallen on a Saturday, there would follow twenty-one days of wind or rain; for

"If the moon on a Saturday be new or full,
There always *was* rain, and there always *wüll*."

One rustic rhyme rehearsed in some places is:—

"A Saturday moon,
If it comes once in seven years,
Comes once too soon."

Next to the day, the medium through which the new moon is first beheld, is of vital moment. In Staffordshire it is unlucky to see this sight through trees. A correspondent in *Notes and Queries* (21st January, 1882) once saw a person almost in tears because she looked on the new moon through her veil, feeling convinced that misfortune would follow. Henderson cites a canon to be observed by those who would know what year they would wed. "Look at the first new moon of the year through a silk handkerchief which has never been washed. As many moons as you see through the handkerchief (the threads multiplying the vision), so many years will pass ere you are married."[404] Hunt tells us, what in fact is widely believed, that "to see the new moon for the first time through glass, is unlucky; you may be certain that you will break glass before that moon is out. I have known persons whose attention has been called to a clear new moon hesitate. 'Hev I seed her out o' doors afore?' if not, they will go into the open air, and, if possible, show the moon 'a piece of gold,' or, at all events, turn their money."[405] Mrs. Latham says: "Many of our Sussex superstitions are probably of Saxon origin; amongst which may be the custom of bowing or curtseying to the new or Lady moon, as she is styled, to deprecate bad luck. There is another kindred superstition, that the Queen of night will dart malignant rays upon you, if on the first day of her re-appearance you look up to her without money in your pocket. But if you are not fortunate enough to have any there, in order to avert her evil aspect, you must immediately turn head over heels! It is considered unlucky to see the new moon through a window-pane, and I have known a maidservant shut her eyes when closing the shutters lest she should unexpectedly see it through the glass. Do not kill your pig until full moon, or the pork will be ruined."[406] In Suffolk, also, "it is considered unlucky to kill a pig in the wane of the moon; if it is done, the pork will waste in boiling. I have known the shrinking of bacon in the pot attributed to the fact of the pig having been killed in the moon's decrease; and I have also known the death of poor piggy delayed, or hastened, so as to happen during its increase."[407]

The desirability of possessing *silver* in the pocket, and of turning it over, when the new moon is first seen, is a point of some interest. Forbes Leslie says, "The ill-luck of having no *silver* money—coins of other metals being of no avail—when you first see or hail a new moon, is still a common belief from Cornwall to Caithness, as well as in Ireland."[408] And Jamieson writes: "Another superstition, equally ridiculous and unaccountable, is still regarded by some. They deem it very unlucky to see the new moon for the first time without having *silver* in one's pocket. Copper is of no avail."[409] We venture to think that this is not altogether unaccountable. The moon at night, in a clear sky, reflects a brilliant whiteness. The two Hebrew words used of this luminary in the Bible, mean "pale light" and "white." "Hindooism says that the moon, Soma, was turned into a female called Chandra—'the White or Silvery One.'"[410] The Santhals of India call the sun *Chando*, which means bright, and is also a name for the moon. Now pure silver is of a very white colour and of a strong metallic lustre. It was one of the earliest known metals, and used as money from the remotest times. Its whiteness led the ancient astrologers, as it afterwards led the alchemists, to connect it with the moon, and to call it Diana and Luna, names previously given to the satellite. For Artemis, the Greek Diana, the Ephesian craftsmen made silver shrines. The moon became the symbol of silver; and to this day fused nitrate of silver is called *lunar* caustic. It was natural and easy for superstition to suppose that silver was the moon's own metal; and to imagine that upon the reappearance of the lunar deity or demon, its beams should be propitiated by some argentine possession. We find that silver was exclusively used in the worship of the moon in Peru.

In a book published in the earlier part of last century, and attributed to Daniel Defoe, we read; "To see a new moon the first time after her change, on the right hand, or directly before you, betokens the utmost good fortune that month; as to have her on your left, or behind you, so that in turning your head back you happen to see her, foreshows the worst; as also, they say, to be without gold in your pocket at that time is of very bad consequence."[411] The mistake in substituting gold for silver here is easily explained. As among the Romans *aes* meant both copper and money; and among the French *argent* means both silver and money in general; so in England gold is the common expression for coin of any substance. Silver being *money*, the word gold was thus substituted; the generic for the specific. Other superstitions besides those above noticed are found in different parts of our enlightened land. Denham says, "I once saw an aged matron turn her apron to the new moon to insure good luck for the ensuing month."[412] And Halliwell mentions a prayer customary among some persons:—

"I see the moon, and the moon sees me.
God bless the moon, and God bless me."[413]

In Devonshire it is lucky to see the new moon over the right, but unlucky to see it over the left shoulder; and to see it straight before is good fortune to the end of the month. "In Renfrewshire, if a man's house be burnt during the wane of the moon, it is deemed unlucky. If the same misfortune take place when the moon is waxing, it is viewed as a presage of prosperity. In Orkney, also, it is reckoned unlucky to flit, or to remove from one habitation to another, during the waning of the moon."[414] A recent writer tells us that in Orkney "there are superstitions likewise associated with the moon. The increase, and full growth, and wane of that satellite are the emblems of a rising, flourishing, and declining fortune. No business of importance is begun during the moon's wane, if even an animal is killed at that period, the flesh is supposed to be unwholesome. A couple to think of marrying at that time would be regarded as recklessly careless respecting their future happiness Old people in some parts of Argyllshire were wont to invoke the Divine blessing on the moon after the monthly change. The Gaelic word for fortune is borrowed from that which denotes the full moon; and a marriage or birth occurring at that period is believed to augur prosperity."[415]

Kirkmichael, says another writer on the Highlands of Scotland, hath "its due proportion of that superstition which generally prevails over the Highlands. Unable to account for the cause, they consider the effects of times and seasons as certain and infallible. The moon in her increase, full growth, and in her wane, are with them the emblems of a rising, flourishing, and declining fortune. At the last period of her revolution they carefully avoid to engage in any business of importance; but the first and the middle they seize with avidity, presaging the most auspicious issue to their undertakings. Poor Martinus Scriblerus never more anxiously watched the blowing of the west wind to secure an heir to his genius, than the love-sick swain and his nymph for the coming of the new moon to be noosed together in matrimony. Should the planet happen to be at the height of her splendour when the ceremony is performed, their future life will be a scene of festivity, and all its paths strewed over with rosebuds of delight. But when her tapering horns are turned towards the north, passion becomes frost-bound, and seldom thaws till the genial season again approaches. From the moon they not only draw prognostications of the weather, but according to their creed also discover future events. There they are clearly portrayed, and ingenious illusion never fails in the explanation. The veneration paid to this planet, and the opinion of its influences, are obvious from the meaning still affixed to some words of the Gaelic language. In Druidic mythology, when the circle of the moon was complete, fortune then promised to be most propitious. Agreeably to this idea, *rath*, which signifies in Gaelic a wheel or circle, is transferred to signify fortune."[416]

Forbes Leslie writes: "The influence which the moon was supposed to exercise on mankind, as well as on inanimate objects, may be traced in the practice of the Druids. It is not yet extinct in Scotland; and the moon, in the increase, at the full, and on the wane, are emblems of prosperity, established success, or declining fortune, by which many persons did, and some still do, regulate the period for commencing their most important undertakings."[417] And yet once more, to make the induction most conclusive; we are told that "the canon law anxiously prohibited observance of the moon as regulating the period of marriage; nor was any regard to be paid to certain days of the year for ceremonies. If the Lucina of the ancients be identified with Diana, it was not unreasonable to court the care of the parturient, by selecting the time deemed most propitious. The strength of the ecclesiastical interdiction does not seem to have prevailed much in Scotland. Friday, which was consecrated to a northern divinity, has been deemed more favourable for the union. In the southern districts of Scotland, and in the Orkney Islands, the inhabitants preferred the increase of the moon for it. Auspicious circumstances were anticipated in other parts, from its celebration at full moon. Good fortune depended so much on the increase of that luminary, that nothing important was undertaken during its wane. Benefit even accrued to the stores provided during its increase, and its effect in preserving them is still credited."[418] To what, but to this prevalent belief in lunar influence on fortune can Shakespeare allude, when Romeo swears:

> "*Rom.* Lady, by yonder blessed moon I swear,
> That tips with silver all these fruit-tree tops—
> *Jul.* Oh, swear not by the moon, the inconstant moon,
> That monthly changes in her circled orb,
> Lest that thy love prove likewise variable."[419]

Upon the physiological influence of the lunar rays in the generation or aggravation of disease, we have but little to add to what has been already written. It is a topic for a special treatise, and properly belongs to those medical experts whose research and practice in this particular branch of physics qualify them to speak with plenary authority. Besides, it has been so wisely handled by Dr. Forbes Winslow, in his admirable monograph on *Light*, that inquirers cannot follow a safer guide than his little book affords. Dr. Winslow accounts for the theory of planetary influence partly by the action of the moon in producing the tides. He says: "Astronomers having admitted that the moon was capable of producing this physical effect upon the waters of the ocean, it was not altogether unnatural that the notion should become not only a generally received but a popular one, that the ebb and flow of the tides had a material influence over the bodily functions. The Spaniards imagine that all who die of chronic diseases breathe their last during the ebb. Southey says, that amongst the wonders of the isles and city of Cadiz, which the historian of that city, Suares de Salazar, enumerates, one is, according to p. Labat, that the sick never die there while the tide is rising or at its height, but always during the ebb. He restricts the notion to the isle of Leon, but implies that the effect was there believed to take place in diseases of all kinds, acute as well as chronic. 'Him fever,' says the negro in the West Indies, 'shall go when the water come low; him always come not when the tide high.' The popular notion amongst the negroes appears to be that the ebb and flow of the tides are caused by a '*fever of the sea*,' which rages for six hours, and then intermits for as many more."[420] Dr. Winslow then subjoins a long list of learned authorities, several of whose writings he subjects to a brief analysis. He disapproves of the presumption that the subject is altogether visionary and utopian; and affirms that it has not always been pursued by competent observers. Periodicity is noted as an important symptom in disease; a feature in febrile disturbance which the present writer himself had abundant opportunity of marking and measuring during an epidemic of yellow fever in the city of Savannah in the year 1876. This periodicity Dr. Winslow regards as the foundation of the alleged lunar influence in morbid conditions. Some remarkable cases are referred to, which, if the fact of the moon's interference with human functions could be admitted, would go a long way to corroborate and confirm it. The supposed influence of the moon on plants is not passed over, nor the chemical composition of lunar light as a possible evil agency. Still considering the matter *sub judice*, Dr. Winslow then proceeds to the alleged influence of the moon on the insane; a question with which he was pre-eminently competent to cope. After alluding to the support given to the popular belief by poets and philosophers of ancient and modern times, the question of periodicity, or "lucid intervals," is again discussed, this time in its mental aspect, and the hygienic or sanatory influence of light is allowed its meed of consideration. The final result of the investigation is that the matter is held to be purely speculative, and it is esteemed wise to hold in reserve any theory in relation to the subject that may have been formed. With this conclusion we are greatly disappointed. Dr. Winslow's aid in the inquiry is most valuable, and if he, after his careful review of pathological literature on lunar influence, coupled with his own extended experience, holds the question in abeyance, who will venture upon a decision? We however believe, notwithstanding every existing difficulty, that the subject will be brought into clear light ere long, and all superstition end in accurate science. Meanwhile, many, even of the enlightened, will cling to the unforgotten fancy which gave rise to the word *lunatic*, and in cases of mental derangement will moralize with young Banks in the *Witch of Edmonton* (1658), "When the moon's in the full, then wit's in the wane."

MOON INHABITATION

Science having practically diminished the moon's distance, and rendered distinct its elevations and depressions, it is natural for "those obstinate questionings of sense and outward things" to urge the inquiry, *Is the moon inhabited*? This question it is easier to ask than to answer. It has been a mooted point for many years, and our wise men of the west seem still disposed to give it up, or, at least, to adjourn its decision for want of evidence. Of "guesses at truth" there have been a great multitude, and of dogmatic assertions not a few; but demonstrations are things which do not yet appear. We now take leave to report progress, and give the subject a little ventilation. We do not expect to furnish an Ariadne's thread, but we may hope to find some indication of the right way out of this labyrinth of uncertainty. *Veritas nihil veretur nisi abscondi*: or, as the German proverb says, "Truth creeps not into corners"; its life is the light.

But before we advance a single step, we desire to preclude all misunderstanding on one point, by distinctly avowing our conviction that the teachings of Christian theology are not at all involved in the issue of this discussion, whatever it may prove. Infinite harm has been done by confusing the religion of science with the science of religion. Religion *is* a science, and science is a religion; but they are not identical. Philosophy ought to be pious, and piety ought to be philosophical; but philosophy and piety are two quantities and qualities that may dwell apart, though, happily, they may also be found in one nature. Each has its own faculties and functions; and in our present investigation, religion has nothing more to do than to shed the influence of reverence, humility, and teachableness over the scientific student as he ponders his problem and works out the truth. In this, and in kindred studies, we may yield without reluctance what a certain professor of religion concedes, and grant without grudging what a certain professor of science demands. Dr. James Martineau says, "In so far as Church belief is still committed to a given kosmogony and natural history of man, it lies open to scientific refutation"; and again, "The whole history of the Genesis of things Religion must unconditionally surrender to the Sciences."[421] In this we willingly concur, for science ought to be, and will be, supreme in its own domain. Bishop Temple does "not hesitate to ascribe to Science a clearer knowledge of the true interpretation of the first chapter of Genesis, and to scientific history a truer knowledge of the great historical prophets. Science enters into Religion, and the believer is bound to recognise its value and make use of its services."[422] Then, to quote the professor of science, Dr. John Tyndall says. "The impregnable position of science may be described in a few words. We claim, and we shall wrest from Theology, the entire domain of cosmological theory."[423] We wish the eloquent professor all success. It was not the spirit of primitive Christianity, but the spirit of priestly ignorance, intolerance, and despotism, which invaded the territory of natural science; and if those who are its rightful lords can recover the soil, we bid them heartily, God speed! We have been driven to these remarks by a twofold impulse. First, we can never forget the injury that has been inflicted on science by the oppositions of a headless religion; any more than we can forget the injury which has been inflicted on religion by the oppositions of a heartless science. Secondly, we have seen this very question of the inhabitation of the planets and satellites rendered a topic of ridicule for Thomas Paine, and an inviting theme for raillery to others of sophistical spirit, by the way in which it has been foolishly mixed up with sacred or spiritual concerns. Surely, the object of God in the creation of our terrestrial race, or the benefits of the death of Jesus Christ, can have no more to do with the habitability of the moon, than the doctrine of the Trinity has to do with the multiplication table and the rule of three, or the hypostatical union with the chemical composition of water and light. Having said thus much of compulsion, we return, not as ministers in the temple of religion so much as students in the school of science, to consider with docility the question in dispute, *Is the moon inhabited*?

Three avenues, more or less umbrageous, are open to us; all of which have been entered. They may be named *observation*, *induction*, and *analogy*. The first, if we could pursue it, would explicate the enigma at once. The second, if clear, would satisfy our reason, which, in such a matter, might be equivalent to sight. And the third might conduct us to a shadow which would "prove the substance true." We begin by dealing briefly with the argument from *observation*. Here our data are small and our difficulties great. One considerable inconvenience in the inquiry is, of course, the moon's distance. Though she is our next-door neighbour in the many-mansioned universe, two hundred and thirty-seven thousand miles are no mere step heavenward. Transit across the intervenient space being at present impracticable, we have to derive our most enlarged views of this "spotty globe" from the "optic glass." But this admirable appliance, much as it has revealed, is thus far wholly inadequate to the solution of our mystery. Robert Hooke, in the seventeenth century, thought that he could construct a telescope with which we might discern the inhabitants of the moon life-size—seeing them as plainly as we see the inhabitants of the earth. But, alas! the sanguine mathematician died in his sleep, and his dream has not yet come true. Since Hooke's day gigantic instruments have been fitted up, furnished with all the modern improvements which could be supplied through the genius or generosity of such astronomers as Joseph Fraunhofer and Sir William Herschel, the third Earl of Rosse and the fourth Duke of Northumberland. But all of these worthy men left something to be done by their successors. Consequently, not long since, our scientists set to work to increase their artificial eyesight. The Rev. Mr. Webb tells us that "the first 'Moon Committee' of the British Association recommended a power of 1,000." But he discourages us if we anticipate large returns; for he adds: "Few indeed are the instruments or the nights that will bear it; but when employed, what will be the result? Since increase of magnifying is equivalent to decrease of distance, we shall see the moon as large (though not as distinct) as if it were 240 miles off, and any one can judge what could be made of the grandest building upon earth at that distance."[424] If therefore we are to see the settlement of the matter in the speculum of a telescope, it may be some time before we have done with what Guillemin calls "the interesting, almost insoluble question, of the existence of living and organized beings on the surface of the satellite of our little earth."[425] Some cynic may interpose with the quotation,—

> "But optics sharp it needs, I ween,
> To see what is not to be seen."[426]

True, but it remains to be shown that there is nothing to be seen beyond what *we* see. We are not prepared to deny the existence of everything which our mortal eyes may fail to trace. Four hundred years ago all Europe believed that to sail in search of a western continent was to wish "to see what is not to be seen"; but a certain Christopher Columbus went out persuaded of things not seen as yet, and having embarked in faith he landed in sight. The lesson must not be lost upon us.

> "There are more things in heaven and earth, Horatio,
> Than are dreamt of in your philosophy."

Because we cannot now make out either habitations or habitants on the moon, it does not necessarily follow that the night will never come when, through some mightier medium than any ever yet constructed or conceived, we shall descry, beside mountains and valleys, also peopled plains and populous cities animating the fair features of this beautiful orb. One valuable auxiliary of the telescope, destined to play an important part in lunar discovery, must not be overlooked. Mr. Norman Lockyer says, "With reference to the moon, if we wish to map her correctly, it is now no longer necessary to depend on ordinary eye observations alone; it is perfectly clear that by means of an image of the moon, taken by photography, we are able to fix many points on the lunar surface."[427] With telescopic and photographic lenses in skilled hands, and a wealth of inventive genius in fertile brains, we can afford to wait a long while before we close the debate with a final negative.

In the meantime, eyes and glasses giving us no satisfaction, we turn to scientific *induction*. Speculation is a kind of mental mirror, that before now has anticipated or supplemented the visions of sense. Not being practical astronomers ourselves, we have to follow the counsel of that unknown authority who bids us believe the expert. But expertness being the fruit of experience, we may be puzzled to tell who have attained that rank. We will inquire, however, with due docility, of the oracles of scientific research. It is agreed on all sides that to render the moon habitable by beings at all akin with our own kind, there must be within or upon that body an atmosphere, water, changing seasons, and the alternations of day and night. We know that changes occur in the moon, from cold to heat, and from darkness to light. But the lunar day is as long as 291 of ours; so that each portion of the surface is exposed to, or turned from, the sun for nearly 14 days. This long exposure produces excessive heat, and the long darkness excessive cold. Such extremities of temperature are unfavourable to the existence of beings at all like those living upon the earth, especially if the moon be without water and atmosphere. As these two desiderata seem indispensable to lunar inhabitation, we may chiefly consider the question, Do these conditions exist? If so, inductive reasoning will lead us to the inference, which subsequent experience will strengthen, that the moon is inhabited like its superior planet. But if not, life on the satellite similar to life on the earth, is altogether improbable, if not absolutely impossible.

The replies given to this query will be by no means unanimous. But, for the full understanding of the state of the main question, and to assist us in arriving at some sort of verdict, we will hear several authorities on both sides of the case. The evidence being cumulative, we pursue the chronological order, and begin with La

Place. He writes: "The lunar atmosphere, if any such exists, is of an extreme rarity, greater even than that which can be produced on the surface of the earth by the best constructed air-pumps. It may be inferred from this that no terrestrial animal could live or respire at the surface of the moon, and that if the moon be inhabited, it must be by animals of another species."[428] This opinion, as Sir David Brewster points out, is not that the moon has no atmosphere, but that if it have any it is extremely attenuated. Mr. Russell Hind's opinion is similar with respect to water. He says: "Earlier selenographists considered the dull, grayish spots to be water, and termed them the lunar seas, bays, and lakes. They are so called to the present day, though we have strong evidence to show that if water exist at all on the moon, it must be in very small quantity."[429] Mr. Grant tells us that "the question whether the moon be surrounded by an atmosphere has been much discussed by astronomers. Various phenomena are capable of indicating such an atmosphere, but, generally speaking, they are found to be unfavourable to its existence, or at all events they lead to the conclusion that it must be very inconsiderable."[430] Humboldt thinks that Schroeter's assumptions of a lunar atmosphere and lunar twilight are refuted, and adds: "If, then, the moon is without any gaseous envelope, the entire absence of any diffused light must cause the heavenly bodies, as seen from thence, to appear projected against a sky *almost black* in the day-time. No undulation of air can there convey sound, song, or speech. The moon, to our imagination, which loves to soar into regions inaccessible to full research, is a desert where silence reigns unbroken."[431] Dr. Lardner considers it proven "that there does not exist upon the moon an atmosphere capable of reflecting light in any sensible degree," and also believes that "the same physical tests which show the non-existence of an atmosphere of air upon the moon are equally conclusive against an atmosphere of vapour."[432] Mr. Breen is more emphatic. He writes: "In the want of water and air, the question as to whether this body is inhabited is no longer equivocal. Its surface resolves itself into a sterile and inhospitable waste, where the lichen which flourishes amidst the frosts and snows of Lapland would quickly wither and die, and where no animal with a drop of blood in its veins could exist."[433] The anonymous author of the Essay on the *Plurality of Worlds* announces that astronomers are agreed to negative our question without dissent. We shall have to manifest his mistake. His words are: "Now this minute examination of the moon's surface being possible, and having been made by many careful and skilful astronomers, what is the conviction which has been conveyed to their minds with regard to the fact of her being the seat of vegetable or animal life? Without exception, it would seem, they have all been led to the belief that the moon is not inhabited; that she is, so far as life and organization are concerned, waste and barren, like the streams of lava or of volcanic ashes on the earth, before any vestige of vegetation has been impressed upon them; or like the sands of Africa, where no blade of grass finds root."[434] Robert Chambers says: "It does not appear that our satellite is provided with an atmosphere of the kind found upon earth; neither is there any appearance of water upon the surface. . . . These characteristics of the moon forbid the idea that it can be at present a theatre of life like the earth, and almost seem to declare that it never can become so."[435] Schoedler's opinion is concurrent with what has preceded. He writes: "According to the most exact observations it appears that the moon has no atmosphere similar to ours, that on its surface there are no great bodies of water like our seas and oceans, so that the existence of water is doubtful. The whole physical condition of the lunar surface must, therefore, be so different from that of our earth, that beings organized as we are could not exist there."[436] Another German author says: "The observations of Fraunhofer (1823), Brewster and Gladstone (1860), Huggins and Miller, as well as Janssen, agree in establishing the complete accordance of the lunar spectrum with that of the sun. In all the various portions of the moon's disk brought under observation, no difference could be perceived in the dark lines of the spectrum, either in respect of their number or relative intensity. From this entire absence of any special absorption lines, it must be concluded that there is no atmosphere in the moon, a conclusion previously arrived at from the circumstance that during an occultation no refraction is perceived on the moon's limb when a star disappears behind the disk."[437] Mr. Nasmyth follows in the same strain. Holding that the moon lacks air, moisture, and temperature, he says, "Taking all these adverse conditions into consideration, we are in every respect justified in concluding that there is no possibility of animal or vegetable life existing on the moon, and that our satellite must therefore be regarded as a barren world."[438] A French astronomer holds a like opinion, saying: "There is nothing to show that the moon possesses an atmosphere; and if there was one, it would be perceptible during the occultations of the stars and the eclipses of the sun. It seems impossible that, in the complete absence of air, the moon can be peopled by beings organized like ourselves, nor is there any sign of vegetation or of any alteration in the state of its surface which can be attributed to a change of seasons."[439] On the same side Mr. Crampton writes most decisively, "With what we *do* know, however, of our satellite, I think the idea of her being inhabited may be dismissed *summarily*; *i.e.* her inhabitation by intelligent beings, or an animal creation such as exist here."[440] And, finally, in one of Maunder's excellent *Treasuries*, we read of the moon, "She has no atmosphere, or at least none of sufficient density to refract the rays of light as they pass through it, and hence there is no water on her surface; consequently she can have no animals like those on our planet, no vegetation, nor any change of seasons."[441] These opinions, recorded by so many judges of approved ability and learning, have great weight; and some may regard their premisses and conclusions as irresistibly cogent and convincing. The case against inhabitation is certainly strong. But justice is impartial. *Audi alteram partem.*

Judges of equal erudition will now speak as respondents. We go back to the seventeenth century, and begin with a work whose reasoning is really remarkable, seeing that it is nearly two hundred and fifty years since it was first published. We refer to the *Discovery of a New World* by John Wilkins, Bishop of Chester; in which the reverend philosopher aims to prove the following propositions:—"1. That the strangeness of this opinion (that the moon may be a world) is no sufficient reason why it should be rejected; because other certain truths have been formerly esteemed as ridiculous, and great absurdities entertained by common consent. 2. That a plurality of worlds does not contradict any principle of reason or faith. 3. That the heavens do not consist of any such pure matter which can privilege them from the like change and corruption, as these inferior bodies are liable unto. 4. That the moon is a solid, compacted, opacous body. 5. That the moon hath not any light of her own. 6. That there is a world in the moon, hath been the direct opinion of many ancient, with some modern mathematicians; and may probably be deduced from the tenets of others. 9. That there are high mountains, deep valleys, and spacious plains in the body of the moon. 10. That there is an atmosphoera, or an orb of gross vaporous air, immediately encompassing the body of the moon. 13. That 'tis probable there may be inhabitants in this other world; but of what kind they are, is uncertain."[442] We go on to 1686, and listen to the French philosopher, Fontenelle, in his Conversations with the Marchioness. "'Well, madam,' *said I*, 'you will not be surprised when you hear that the moon is an earth too, and that she is inhabited as ours is.' 'I confess,' *said she*, 'I have often heard talk of the world in the moon, but I always looked upon it as visionary and mere fancy.' 'And it may be so still,' *said I*. 'I am in this case as people in a civil war, where the uncertainty of what may happen makes them hold intelligence with the opposite party; for though I verily believe the moon is inhabited, I live civilly with those who do not believe it; and I am still ready to embrace the prevailing opinion. But till the unbelievers have a more considerable advantage, I am for the people in the moon.'"[443] Whatever may be thought of his philosophy, no one could quarrel with the Secretary of the Academy on the score of his politeness or his prudence. A more recent and more reliable authority appears in Sir David Brewster. He tells us that "MM. Mädler and Beer, who have studied the moon's surface more diligently than any of their predecessors or contemporaries, have arrived at the conclusion that she has an atmosphere." Sir David himself maintains that "*every planet and satellite in the solar system must have an atmosphere.*"[444] Bonnycastle, whilom professor of mathematics in the Royal Military Academy, Woolwich, writes: "Astronomers were formerly of opinion that the moon had no atmosphere, on account of her never being obscured by clouds or vapours; and because the fixed stars, at the time of an occultation, disappear behind her instantaneously, without any gradual diminution of their light. But if we consider the effects of her days and nights, which are near thirty times as long as with us, it may be readily conceived that the phenomena of vapours and meteors must be very different. And besides, the vaporous or obscure part of our atmosphere is only about the one thousand nine hundred and eightieth part of the earth's diameter, as is evident from observing the clouds, which are seldom above three or four miles high; and therefore, as the moon's apparent diameter is only about thirty-one minutes and a half, or one thousand eight hundred and ninety seconds, the obscure part of her atmosphere, supposing it to resemble our own, when viewed from the earth, must subtend an angle of less than one second; which is so small a space, that observations must be extremely accurate to determine whether the supposed obscuration takes place or not."[445] Dr. Brinkley, at one time the Astronomer-Royal of Ireland, writes: "Many astronomers formerly denied the existence of an atmosphere at the moon; principally from observing no variation of appearance on the surface, like what would take place, did clouds exist as with us; and also, from observing no change in the light of the fixed stars on the approach of the dark edge of the moon. The circumstance of there being no clouds, proves either that there is no atmosphere similar to that of our earth, or that there are no waters on its surface to be converted into vapour; and that of the lustre of the stars not being changed, proves that there can be no dense atmosphere. But astronomers now seem agreed that an atmosphere does surround the moon, although of small density when compared with that of our earth. M. Schroeter has observed a small twilight in the moon, such as would arise from an atmosphere capable of reflecting the rays at the height of about one mile."[446] Dr. Brinkley is inaccurate in saying that astronomers are agreed as to the lunar atmosphere. Like students in every other department of inquiry, spiritual as well as physical, they fail at present to see "eye to eye"; which is not surprising, seeing that the eye is so restricted, and the object so remote.

Dr. Dick, whose productions have done much to popularize the study of the heavens, and to promote its reverent pursuit, says: "On the whole it appears most probable that the moon is surrounded with a fluid which serves the purpose of an atmosphere; although this atmosphere, as to its nature, composition, and refractive power, may be very different from the atmosphere which surrounds the earth. It forms no proof that the moon, or any of the planets, is destitute of an atmosphere, because its constitution, its density, and its power of refracting the rays of light are different from ours. An atmosphere may surround a planetary body, and yet its parts be so fine and transparent that the rays of light, from a star or any other body, may pass through it without being in the least obscured, or changing their direction. In our reasonings on this subject, we too frequently proceed on the false principle, that everything connected with other worlds must bear a resemblance to those on the earth."[447] Mr. Neison, who has written one of the latest contributions to the science of selenography, says, "Of the present non-existence of masses of water upon the surface of the moon, there remains no doubt, though no evidence of its entire absence from the lunar crust can be adduced; and similarly, many well-established facts in reference to the moon afford ample proof of the non-existence of a lunar atmosphere, having a density equal to, or even much less than, that of the earth; but of the absence of an atmosphere, whose mass should enable it to play an important part in the moulding of the surface of the moon, and comparable almost to that of the terrestrial atmosphere, in their respective ratios to the masses of their planets, little, if any, trustworthy evidence exists." On another page of the same work, the author affirms "that later inquiries have shown that the moon may possess an atmosphere that must be regarded as fully capable of sustaining various forms of vegetation of even an advanced type; and, moreover, it does not appear how it can justly be questioned that the lunar surface in favourable positions may yet retain a sufficiency of moisture to support vegetation of many kinds; whilst in a very considerable portion of the entire surface of the moon, the temperature would not vary sufficiently to materially affect the existence of vegetable life."[448] Some of these writers may appear to be travelling rather too fast or too far, and their assumptions may wear more of the aspect of plausibility than of probability. But on their atmospheric and aqueous hypothesis, vegetation in abundance is confessedly a legitimate consequence. If a recent writer has liberty to condense into a sentence the conclusion from the negative premiss in the argument by saying, "As there is but a little appearance of water or air upon the moon, the conclusion has been inferred that there exists no vegetable or animal life on that globe,"[449] other writers, holding opposite views of the moon's physical condition, may be allowed to expatiate on the luxuriant life which an atmosphere with water and temperature would undoubtedly produce. Mr. Proctor's tone is temperate, and his language that of one who is conscious with Hippocrates that "art is long and life is short." He says, in one of his contributions to lunar science, "It may safely be asserted that the opportunities presented during the life of any single astronomer for a trustworthy investigation of any portion of the moon's surface, under like conditions, are few and far between, and the whole time so employed must be brief, even though the astronomer devote many more years than usual to observational research."[450] This prepares us to find in another of the same author's works the following suggestive sentence: "With regard to the present habitability of the moon, it may be remarked that we are not justified in asserting positively that no life exists upon her surface. Life has been found under conditions so strange, we have been so often mistaken in assuming that *here* certainly, or *there*, no living creatures can possibly exist, that it would be rash indeed to dogmatise respecting the state of the moon in this respect."[451] Narrien, one of the historians of the science, may be heard, though his contribution might be cast into either scale. He writes: "The absence of those variations of light and shade which would be produced by clouds floating above her surface, and the irregularities of the ground, visible at the bottom and on the sides of her cavities, have given reason to believe that no atmosphere surrounds her, and that she is destitute of rivers and seas. Such are the opinions generally entertained concerning the moon; but M. Schroeter, a German astronomer, ventures to assert that our satellite is the abode of living and intellectual beings; he has perceived some indications of an atmosphere which, however, he admits, cannot exceed two miles in height, and certain elevations which appear to him to be works of art rather than of nature. He considers that a uniformity of temperature must be produced on her surface by her slow rotation on her axis, by the insensible change from day to night, and the attenuated state of her atmosphere, which is never disturbed by storms; and that light vapours, rising from her valleys, fall in the manner of a gentle and refreshing dew to fertilize her fields."[452] Dr. H. W. M. Olbers is fully persuaded "that the moon is inhabited by rational creatures, and that its surface is more or less covered with a vegetation not very dissimilar to that of our own earth." Dr. Gruithuisen, of Munich, maintains that he has descried through his large achromatic telescope "great artificial works in the moon erected by the lunarians," which he considers to be "a system of fortifications thrown up by the selenitic engineers." We should have scant hope of deciding the dispute by the dicta of the ancients, were these far more copious than we find them to be. Yet reverence for antiquity may justify our quoting one of the classic fathers. Plutarch says, "The Pythagoreans affirme, that the moone appeereth terrestriall, for that she is inhabited round about, like as the earth wherein we are, and peopled as it were with the greatest living creatures, and the fairest plants." Again, "And of all this that hath been said (my friend *Theon*) there is nothing that doth proove and show directly, this habitation of men in the moon to be impossible."[453] Here we close the argument based on *induction*, and sum up the evidence in our possession. On the one hand, several scientific men, whose names we need not repeat, having surveyed the moon, deny it an atmosphere, water, and other conditions of life. Consequently, they disbelieve in its inhabitation, solely because they consider the fact undemonstrable; none of them being so unscientific as to believe it to be absolutely impossible. On the other hand, we have the valuable views of Mädler and Beer, whose lunar labours are unsurpassed, and whose map of the moon is a marvel and model of advanced selenography. They do not suppose the conditions on our satellite to be exactly what they are on this globe. In their own words, the moon is "no copy of the earth, much less a colony of the same." They merely believe her to be environed with air, and thus habitable. And when we recall our own Sir David Brewster, Professor Bonnycastle, Dr. Brinkley, Dr. Dick, Mr. Neison, and Mr. Proctor; and reckon with them the continental astronomers, Dr. Gruithuisen, Dr. Olbers, and Schroeter, all of whom attempted to fix the idea of planetary inhabitation on the popular mind, we must acknowledge that they, with their opponents, have a strong claim on our attention. The only verdict we are able just now to render, after hearing these conflicting testimonies, is the Scotch one, *Not proven*. We but append the legal indorsement *ignoramus*, we do not know. The subject must remain *sub judice*; but what we know not now, we hope to know hereafter.

Having interrogated *sense* and *science*, with the solution of our enigma anything but complete, we resort last of all to the argument from *analogy*. If this can illumine the obscurity, it will all be on the positive side of the inquiry. At present the question resembles a half-moon: analogy may show that the affirmative is waxing towards a full-orbed conviction. We open with Huyghens, a Dutch astronomer of note, who, while he thinks it certain "that the moon has no air or atmosphere surrounding it as we have," and "cannot imagine how any plants or animals whose whole nourishment comes from fluid bodies, can thrive in a dry, waterless, parched soil," yet asks, "What, then, shall this great ball be made for; nothing but to give us a little weak light in the night time, or to raise our tides in the sea? Shall not we plant some people there that may have the pleasure of seeing our earth turn upon its axis, presenting them sometimes with a prospect of Europe and Africa, and then of Asia and America; sometimes half and sometimes full?"[454] Ray was "persuaded that this luminary doth serve many ends and uses, especially to maintain the creatures which in all likelihood breed and inhabit there."[455] Swedenborg's *ipse dixit* ought to convince the most incredulous; for he speaks "from what has been heard and seen." Thus he says: "That there are inhabitants in the moon is well known to spirits and angels, and in like manner that there are inhabitants in the moons or satellites which revolve about Jupiter and Saturn. They who have not seen and discoursed with spirits coming from those moons still entertain no doubt but there are men inhabiting them, because they are earths alike with the planets, and wherever an earth is, there are men inhabitants; for man is the end for which every earth was created, and nothing was made by the great Creator without an end."[456] If any are still sceptical, Sir William Herschel, an intellectual light of no mean magnitude, may reach them. He writes: "While man walks upon the ground, the birds fly in the air, and fishes swim in water, we can certainly not object to the conveniences afforded by the moon, if those that are to inhabit its regions are fitted to their conditions as well as we on this globe arc to ours. An absolute or total sameness seems rather to denote imperfections, such as nature never exposes to our view; and, on this account, I believe the analogies that have been mentioned fully sufficient to establish the high probability of the moon's being inhabited like the earth."[457] The voice of Dr. Dwight, the American theologian, will not be out of harmony here. In discoursing of the starry heavens, he says of the planets: "Of these inferior worlds, the moon is one; and to us, far the most interesting. How many important purposes which are known does this beautiful attendant of our earth continually accomplish! How many more, in all probability, which are hitherto unknown, and which hereafter may be extensively disclosed to more enlightened, virtuous, and happy generations of men! At the same time, it is most rationally concluded that intelligent beings in great multitudes inhabit her lucid regions, being far better and happier than ourselves."[458] Whewell's *Bridgewater Treatise* will furnish us a fitting quotation. "The earth, the globular body thus covered with life, is not the only globe in the universe. There are, circling about our own sun, six others, so far as we can judge, perfectly analogous in their nature: besides our moon and other bodies analogous to it. No one can resist the temptation to conjecture, that these globes, some of them much larger than our own, are not dead and barren:—that they are, like ours, occupied with organization, life, intelligence."[459] In a most eloquent passage, Dr. Chalmers, who will always be heard with admiration, exclaims: "Who shall assign a limit to the discoveries of future ages? Who shall prescribe to science her boundaries, or restrain the active and insatiable curiosity of man within the circle of his present acquirements? We may guess with plausibility what we cannot anticipate with confidence. The day may yet be coming when our instruments of observation shall be inconceivably more powerful. They may ascertain still more decisive points of resemblance. They may resolve the same question by the evidence of sense which is now so abundantly convincing by the evidence of analogy. They may lay open to us the unquestionable vestiges of art, and industry, and intelligence. We may see summer throwing its green mantle over those mighty tracts, and we may see them left naked and colourless after the flush of vegetation has disappeared. In the progress of years or of centuries, we may trace the hand of cultivation spreading a new aspect over some portion of a planetary surface. Perhaps some large city, the metropolis of a mighty empire, may expand into a visible spot by the powers of some future telescope. Perhaps the glass of some observer, in a distant age, may enable him to construct the map of another world, and to lay down the surface of it in all its minute and topical varieties. But there is no end of conjecture; and to the men of other times we leave the full assurance of what we can assert with the highest probability, that yon planetary orbs are so many worlds, that they teem with life, and that the mighty Being who presides in high authority over this scene of grandeur and astonishment has there planted the worshippers of His glory."[460]

How fine is this outburst of the great Scotch orator! He spoke as one inspired with prophetic foreknowledge; for in less than twenty years after this utterance, Beer and Mädler published their splendid *Mappe Selenographica*, or map of the moon; and photography offered its aid to the fuller delineation of our silvery satellite. Who can tell what the last fifteen years of this eventful century may develop in the same direction? Verily these intuitions of reason seem often favoured with an apocalypse of coming disclosures; and, if we may venture to adopt with slight alteration a sentence of Shelley, we will say: "It is impossible to read the compositions of the most celebrated writers of the present day without being startled with the electric life which burns within their words. They measure the circumference and sound the depths of nature with a comprehensive and all-penetrating spirit, and they are themselves perhaps the most sincerely astonished at its manifestations; for it is less their spirit than the spirit of the age." The poets of science, in their analogies, are "the hierophants of an unapprehended inspiration; the mirrors of the gigantic shadows which futurity casts upon the present."[461] Equally noble with the language of Chalmers is a paragraph which we have extracted from a work by that scholarly writer, Isaac Taylor. He says: "There are two facts, each of which is significant in relation to our present subject, and of which the first has long been understood, while the latter (only of late ascertained) is every day receiving new illustrations; namely, that our planet is, in no sense, of primary importance in the general system, or entitled, by its magnitude, or its position, or its constitution, to be considered as exerting any peculiar influence over others, or as the object of more regard than any others. This knowledge of our real place and value in the universe is a very important consequence of our modern astronomy, and should not be lost sight of in any of our speculations. But then it is also now ascertained that the great laws of our own planet, and of the solar system to which it belongs, prevail in all other and the most remote systems, so as to make the visible universe, in the strictest sense, ONE SYSTEM—indicating one origin and showing the presence of one Controlling Power. Thus the law of gravitation, with all the conditions it implies, and the laws of light, are demonstrated to be in operation in regions incalculably remote; and just so far as the physical constitution of the other planets of our system can be either traced, or reasonably conjectured, it appears that, amid great diversities of constitution, the same great principles prevail in all; and therefore our further conjecture concerning the existence of sentient and rational life in other worlds is borne out by every sort of analogy, abstract and physical; and this same rule of analogy impels us to suppose that rational and moral agents, in whatever world found, and whatever diversity of form may distinguish them, would be such that we should soon feel at home in their society, and able to confer with them, to communicate knowledge to them, and to receive knowledge from them. Neither truth nor virtue is local; nor can there be wisdom and goodness in one planet, which is not wisdom and goodness in every other."[462] The writer of the *Plurality of Worlds*, a little work distinct from the essay already quoted, vigorously vindicates "the deeply cherished belief of some philosophers, and of many Christians, that our world, in its present state, contains the mere embryo of intelligent, moral, and religious happiness; that the progress of man in his present state is but the initiation of an interminable career of glory; and that his most widely extended associations are a preparation for as interminably an intercourse with the whole family of an intelligent universe."[463] Dr. Arnott may add a final word, a last link in this evidential chain of analogy. He writes: "To think, as our remote forefathers did, that the wondrous array of the many planets visible from this earth serve no purpose but to adorn its nocturnal sky, would now appear absurd indeed; but whether they are inhabited by beings at all resembling the men of this earth, we have not the means of knowing. All the analogies favour the opinion that they are the abodes of life and its satisfactions. On this earth there is no place so hot or so cold, so illumined or so dark, so dry or so wet, but that it has creatures constituted to enjoy life there."[464]

Here our long list of learned authorities shall terminate. We have strung together a large number of citations, and have ourselves furnished only the string. Indeed, what more have amateurs that they can do? For, as Pope puts it,—

> "Who shall decide, when doctors disagree,
> And soundest casuists doubt, like you and me?"

Besides, astronomy is no child's play, nor are its abstruse problems to be mastered by superficial meddlers. "Its intricacy," as Narrien reminds us, "in the higher departments, is such as to render the processes unintelligible to all but the few distinguished persons who, by nature and profound application to the subject, are qualified for such researches."[465] But if professionals must be summoned as witnesses, ordinary men may sit as jurors. This function we have wished to fufil; and we avow ourselves considerably perplexed, though not in despair. We hoped that after a somewhat exhaustive examination, we might be able to state the result with an emphasis of conviction. This we find impossible; but we can affirm on which side the evidence appears to preponderate, and whither, we rest assured, further light will lead our willing feet. The conclusion, therefore, of the whole matter is: we cannot see any living creatures on the moon, however long we strain our eyes. No instrument has yet been constructed that will reveal the slightest vestige of inhabitation. Consequently, the actual evidence of sense is all against us, and we resign it without demur. This point, being settled, is dismissed.

Next, we reconsider the results of scientific study, and are strongly inclined to think the weight of testimony favours the existence of a thin atmosphere, at least some water, and a measure of light and shade in succession. These conditions must enable vegetables and animals to exist upon its surface, though their constitution is in all probability not analogous with that of those which are found upon our earth. But to deny the being of inhabitants of some kind, even in the absence of these conditions, we submit would be unphilosophical, seeing that the Power which adapted terrestrial life to terrestrial environments could also adapt lunar life to the environments in the moon. We are seeking no shelter in the miraculous, nor do we run from a dilemma to the refuges of religion. Apart from our theological belief in the potency of the Creator and Controller of all worlds, we simply regard it as illogical and inconclusive to argue that because organization, life, and intelligence obtain within one sphere under one order of circumstances, *therefore* the same order obtains in every other sphere throughout the system to which that one belongs. The unity of nature is as clear to us as the unity of God; but unity is not uniformity. We view the whole creation as we view this world; the entire empire as we view this single province,

> "Where order in variety we see,
> And where, though all things differ, all agree."

And, finally, as analogy is unreservedly on the side of the occupation of every domain in creation, by some creatures who have the dominion, we cannot admit the probability that the earth is the only tenement with tenants: we must be confirmed in our judgment that the sun and the planets, with their moons, ours of course included, are neither blank nor barren, but abodes of variously organized beings, fitted to fulfil the chief end of all noble existence: the enjoyment of life, the effluence of love, the good of all around and the glory of God above.

This article, that the moon is inhabited, may therefore form a clause of our scientific creed; not to be held at any hazard, as a matter of life or death, or a test of communion, but to be maintained subject to corrections such as future elucidation may require. We believe that we are justified by science, reason, and analogy; and confidently look to be further justified by verification. We accept many things as matters of faith, which we have not fully ascertained to be matters of fact; but "faith is the assurance of things hoped for, the proving of things not seen." By double entry the books of science are kept, by reasoning and demonstration: when future auditors shall examine the accounts of the moon's inhabitation, we are persuaded that the result of our reckoning will be found to be correct.

If any would charge us with a wish to be wise above what is written, we merely reply: There are unwritten revelations which are nevertheless true. Besides, we are not sure that at least an intimation of other races than those of the earth is not already on record. Not to prove any position, but to check obstructive criticism, we refer to the divine who is said to have witnessed in magnificent apocalypse some closing scenes of the human drama. If he also heard in sublime oratorio a prelude of this widely extended glory, our vision may not be a "baseless fabric." After the quartettes of earth, and the interludes of angels, came the grand finale, when every creature which is in heaven, as well as on the earth, was heard ascribing "Blessing and honour and glory and power to Him who sitteth upon the throne." Assuredly, our conception of a choir worthy to render that chorus is not of an elect handful of "saints," or contracted souls, embraced within any Calvinistic covenant, but of an innumerable multitude of ennobled, purified, and expanded beings, convoked from every satellite and planet, every sun and star, and overflowing with gratitude and love to that universal Father of lights, with whom is no parallax, nor descension, and who kindled every spark of life and beauty that in their individual and combined lustre He might reflect and repeat His own ineffable blessedness.

APPENDIX

Literature of the Lunar Man.
Vide p. 8.

1. *The Man in the Moone.* Telling Strange Fortunes. London, 1609.

2. "*The Man in the Moone*, discovering a world of Knavery under the Sunne; both in the *Parliament*, the *Councel* of *State*, the *Army*, the *City*, and the *Country*." Dated, "Die Lunae, From Nov. 14 to Wednesday Novemb. 21 1649." *Periodical Publications, London.* British Museum. Another Edition, "Printed for Charles Tyns, at the Three Cups on London Bridge, 1657."

3. "ΣΕΛΗΝΑΡΧΙΑ, *or the Government of the World in the Moon.*" A comical history written by Cyrano Bergerac, and done into English by Tho. St. Serf. London 1659."

The same, Englished by A. Lovell, A.M., London, 1687.

4. "*The Man in the Moon, or Travels into the Lunar Regions*," by W. Thomson, London, 1783.

In this lucubration the Man in the Moon shows the Man of the People (Charles Fox), many eminent contemporaries, by means of a magical glass.

5. "*The Man in the Moon*, consisting of Essays and Critiques." London, 1804. Of no value. After shining feebly like a rushlight for about two months, it went out in smoke.

6. *The Man in the Moon.* London, 1820. A Political Squib.

7. *The Loyal Man in the Moon*, 1820, is a Political Satire, with thirteen cuts.

8. *The Man in the Moon*, London, 1827(?). A Poem. *N.B.* The word *poem* has many meanings.

9. *The Man in the Moon.* Edinburgh, 1832. A small sheet, sold

for political purposes, at the high price of a penny. The Lunar Man pledges himself to "do as I like, and not to care one straw for the opinion of any person on earth."

10. *The Man in the Moon*. London, 1847. This is a comical serial, edited by Albert Smith and Angus B. Reach; and is rich, racy, and now rare.

11. *The Moon's Histories*. By a Lady. London, 1848.

The Mirror of Pythagoras

Vide p. 147.

"In laying thus the blame upon the moone,
Thou imitat'st subtill *Pythagoras*,
Who, what he would the people should beleeve,
The same be wrote with blood upon a glasse,
And turn'd it opposite 'gainst the new moone
Whose beames reflecting on it with full force,
Shew'd all those lynes, to them that stood behinde,
Most playnly writ in circle of the moone;
And then he said, Not I, but the new moone
Fair *Cynthia*, perswades you this and that."

Summer to Sol, in *A Pleasant Comedie, called Summer's Last Will and Testament*. Written by Thomas Nash. London, 1600.

The East Coast of Greenland.
Vide p. 171.

"When an eclipse of the moon takes place, they attribute it to the moon's going into their houses, and peeping into every nook and corner, in search of skins and eatables, and on such occasions accordingly, they conceal all they can, and make as much noise as possible, in order to frighten away their unbidden guest."—*Narrative of an Expedition to the East Coast of Greenland*: Capt. W. A. Graah, of the Danish Roy. Navy. London, 1837, p. 124.

Lord Iddesleigh on the Moon.
Vide p. 189.

Speaking at a political meeting in Aberdeen, on the 22nd of September, 1885, the Earl of Iddesleigh approved the superannuated notion of lunar influence, and likened the leading opponents of his party to the old and new moon. "What signs of bad weather are there which sometimes you notice when storms are coming on? It always seems to me that the worst sign of bad weather is when you see what is called the new moon with the old moon in its arms. I have no doubt that many of you Aberdeen men have read the fine old ballad of Sir Patrick Spens, who was drowned some twenty or thirty miles off the coast of Aberdeen. In that ballad he was cautioned not to go to sea, because his faithful and weatherwise attendant had noticed the new moon with the old moon in its lap. I think myself that that is a very dangerous sign, and when I see Mr. Chamberlain, the new moon, with Mr. Gladstone, the old one, in his arms, I think it is time to look out for squally weather."—*The Standard*, London, Sept. 23rd, 1885.

The Scottish ballad of Sir Patrick Spens, which is given in the collections of Thomas Percy, Sir Walter Scott, William Motherwell, and others, is supposed by Scott to refer to a voyage that may really have taken place for the purpose of bringing back the Maid of Norway, Margaret, daughter of Alexander III., to her own kingdom of Scotland. Finlay regards it as of more modern date. Chambers suspects Lady Wardlaw of the authorship. While William Allingham counsels his readers to cease troubling themselves with the historical connection of this and all other ballads, and to enjoy rather than investigate. Coleridge calls Sir Patrick Spens a "grand old ballad."

Greeting the New Moon in Fiji.

Vide p. 212.

"There is, I find, in Colo ('the devil's country' as it is called), in the mountainous interior of Viti Levu, the largest island of Fiji, a very curious method of greeting the new moon, that may not, as few Europeans have visited this wild part, have been noticed. The native, on seeing the thin crescent rise above the hills, salutes it with a prolonged 'Ah!' at the same time quickly tapping his open mouth with his hand, thus producing a rapid vibratory sound. I inquired of a chief in the town the meaning and origin of this custom, and my interpreter told me that he said, 'We always look and hunt for the moon in the sky, and when it comes we do so to show our pleasure at finding it again. I don't know the meaning of it; our fathers always did so.'"—Alfred St. Johnston, in *Notes and Queries* for July 23rd, 1881, p. 67. See also Mr. St. Johnston's *Camping Among Cannibals*, London, 1883, p. 283.

Lunar Influence on Dreams.
Vide p. 214.

Vide p. 214.

Arnason says that in Iceland "there are great differences between a dream dreamt in a crescent moon, and one dreamt when the moon is waning. Dreams that are dreamt before full moon are but a short while in coming true; those dreamt later take a longer time for their fulfilment."—*Icelandic Legends*, Introductory Essay, p. lxxxvii.

NOTES

1 *The Martyrs of Science*, by Sir David Brewster, K.H., D.C.L. London, 1867, p. 21.

2 *The Marvels of the Heavens*, by Camile Flammarion. London, 1870, p. 238.

3 *The Jest Book*. Arranged by Mark Lemon. London, 1864, p. 310.

4 *Timon*, a Play. Edited by the Rev. A. Dyce. London (Shakespeare Society), 1842, Act iv. Scene iii.

5 *The Man in the Moon drinks Claret*, as it was lately sung at the Court in Holy-well. *Bagford Ballads*, Folio Collection in the British Museum, vol. ii. No. 119.

6 *Conceits, Clinches, Flashes, and Whimzies*. Edited by J. O. Halliwell, F.R.S. London, 1860, p. 41.

7 *The Man in the Moon*, by C. Sloman. London, 1848, Music by E. J. Loder.

8 *Ancient Songs and Ballads*, by Joseph Ritson. London, 1877, p. 58.

9 *On the Religions of India*. Hibbert Lectures for 1878. London, p. 132.

10 *An Etymological Dictionary of the Scottish Language*, by John Jamieson, D.D. Paisley, 1880, iii. 299.

11 *Sir Thomas Browne's Works*. Edited by Simon Wilkin, F.L.S., London, 1835, iii. 157.

12 *Popular Antiquities of Great Britain*. Hazlitt's Edition. London, 1870, ii. 275.

13 *Asgard and the Gods*. Adapted from the work of Dr. Wägner, by M.W. Macdowall; and edited by W. S. Anson. London, 1884, p. 30.

14 *An Introduction to the Science of Comparative Mythology and Folk Lore*, by the Rev. Sir George W. Cox, Bart., M.A. London, 1881, p. 12.

15 *Plutarch's Morals*. Translated by p. Holland. London, 1603, p. 1160.

16 *Myths and Marvels of Astronomy*, by R. A. Proctor. London, 1878, p. 245. See also, *As Pretty as Seven and other German Tales*, by Ludwig Bechstein. London, p. 111.

17 *Curious Myths of the Middle Ages*, by S. Baring-Gould, M.A. London, 1877, p. 193.

18 *Northern Mythology*, by Benjamin Thorpe. London, 1851, iii. 57.

19 *Notes and Queries*. First Series, 1852, vol. vi. p. 232. The entire text of this poem is given in Bunsen's *God in History*. London, 1868, ii. 495.

20 Thorpe's *Mythology*, i. 6.

21 *Ibid.*, 143.

22 *Curious Myths*, pp. 201-203.

23 *Teutonic Mythology*, by Jacob Grimm. Translated by J. S. Stallybrass. London, 1883, ii. 717.

24 *De Natura Rerum*. MS. Harl. No. 3737.

25 MS. Harl. No. 2253, 81.

26 *The Archaeological Journal* for March, 1848, pp. 66, 67.

27 See Tyrwhitt's *Chaucer*. London, 1843, p. 448.

28 Dekker's *Dramatic Works*. Reprinted, London, 1873, ii. 121.

29 *Popular Rhymes of Scotland*. Robert Chambers. London and Edinburgh, 1870, p. 185.

30 *Popular Rhymes and Nursery Tales*, by J. O. Halliwell. London, 1849, p. 228.

31 *Curious Myths*, p. 197.

32 Grimm's *Teutonic Mythology*, ii. 719-20.

33 *The Vision of Dante Alighieri*. Translated by the Rev. H. F. Cary, A.M. London.

34 *The Folk-Lore of China*, by N. B. Dennys, Ph.D. London and Hong Kong, 1876, p. 117.

35 *Himalayan Journals*, by Joseph D. Hooker, M.D., R.N., F.R.S. London, 1855, ii. 278.

36 *Primitive Culture*, by Edward B. Tyler. London, 1871, i 320.

37 *A Brief Account of Bushman Folk-Lore*, by W. H. J. Bleek, Ph.D. Cape Town, 1875, p. 9.

38 *The History of Greenland*, from the German of David Cranz. London, 1820, i. 212.

39 *An Arctic Boat Journey in the Autumn of 1854*, by Isaac J. Hayes, M.D. Boston, U.S., 1883, p. 254.

40 *The Natural Genesis*, by Gerald Massey. London, 1883, i. 115.

41 *The Church Missionary Intelligencer* for November, 1858, p. 249.

42 *Ibid.*, for April, 1865, p. 116.

43 See *Notes and Queries*. First Series. Vol. xi. p. 493.

44 *Researches into the Early History of Mankind*, by Edward B. Tylor, D.C.L., LL.D., F.R.S. London, 1878, p. 378.

45 *Ibid.*, p. 336.

46 *Notes and Queries: on China and Japan*. Hong Kong, August, 1869, p, 123.

47 *Selected Essays on Language, Mythology, and Religion*. London, 1881, i. 613.

48 *Vico*, by Robert Flint. Edinburgh, 1884, p. 210.

49 *The Dictionary Historical and Critical* of Mr. Peter Bayle. London, 1734, v. 576.

50 See *Lunar World*, by the Rev. J. Crampton, M.A. Edinburgh, 1863, p. 83.

51 *Dictionary of Phrase and Fable*, by the Rev. E. Cobham Brewer, LL.D. London, p. 592.

52 *The Man in the Moon*. By an Undergraduate of Worcester College. Oxford, 1839, Part i. p. 3.

53 MS. in the British Museum Library. Additional MSS. No. 11,812.

54 Lucian's *Works*. Translated from the Greek by Ferrand Spence. London, 1684, ii. 182.

55 *The Table Book*. By William Hone. London, 1838, ii. 252.

56 *Adventures of Baron Munchausen*. London, 1809, p. 44.

57 Flammarion's *Marvels of the Heavens*, p. 241.

58 *Records of the Past*. Edited by S. Birch, LL.D., D.C.L. London, iv. 121.

59 *The Philosophie*, 1603, Holland's Transl. p. 1184.

60 *Primitive Culture*, ii. 64.

61 *A Journey to the Moon*, by the Author of *Worlds Displayed*. London, p. 6.

62 Dennys' *Folk-Lore of China*, p. 101.

63 Grimm's *Teutonic Mythology*, ii. 720.

64 Flammarion's *Marvels of the Heavens*, p. 253.

65 *The Philosophie*, p. 338.

66 *The Woman in the Moone*, by John Lyllie. London, 1597.

67 Dr. Rae, *On the Esquimaux*. Transactions of the Ethnological Society, vol. iv., p. 147.

68 *Vide* also *A Description of Greenland*, by Hans Egede. Second Edition. London, 1818, p. 206.

69 *Amazonian Tortoise Myths*, by Ch. Fred. Hartt, A.M. Rio de Janeiro, 1875, p. 40.

70 *Algic Researches*, by Henry Rowe Schoolcraft. New York, 1839, ii. 54.

71 *Information respecting the History, &c., of the Indian Tribes*, by H. R. Schoolcraft. Philadelphia, v. 417.

72 *Nineteen Years in Polynesia*, by the Rev. George Turner. London, 1861, p. 247.

73 *An Account of the Natives of the Tonga Islands in the South Pacific Ocean*, by William Mariner. Arranged by John Martin, M.D. London, 1818, ii. 127.

74 *Myths and Songs from the South Pacific*, by the Rev. W. W. Gill, B.A. London, 1876, p. 45.

75 Grimm's *Teutonic Mythology*, ii. 716.

76 *Selected Essays*, vol. i. note to p. 611.

77 *The Sacred and Historical Books of Ceylon*, edited by Edward Upham. London, 1833, iii. 309.

78 *Teutonic Mythology*, ii. 716.

79 *Illustrations of Shakespeare*. London, 1807, i. 17.

80 *Dictionnaire Infernal*, par J. Collin de Plancy. Paris, 1863, p. 592.

81 *The Chinese Reader's Manual*, by W. F. Mayers. Shanghai, 1874, p. 219.

82 *The Chinese Readers Manual*, p. 95.

83 *Reynard the Fox in South Africa; or, Hottentot Fables and Tales* by W. H. J. Bleek. London, 1864, p. 72.

84 *A Brief Account of Bushman Folk-Lore*, by Dr. Bleek. Cape Town, 1875, p. 10.

85 *Outlines of Physiology, Human and Comparative*, by John Marshall, F.R.S. London, 1867, ii. 625.

86 *Lectures on the Native Regions of Mexico and Peru*, by Albert Réville, D.D. London, 1884, p. 8.

87 *History of the Conquest of Mexico*, by William H. Prescott. London, 1854, p. 50.

88 *The Native Races of the Pacific States of North America*, by Hubert Howe Bancroft. New York, 1875, iii. 62.

89 *Zoological Mythology; or, the Legends of Animals*, by Angelo de Gubernatis. London, 1872, ii. 80.

90 *Ibid.*, ii. 76.

91 *Report on the Indian Tribes Inhabiting the Country in the Vicinity of the 49th Parallel of North Latitude*, by Capt. Wilson. Trans. of Ethnolog. Society of London, 1866. New Series, iv. 304.

92 *The Races of Mankind*, by Robert Brown, M.A., Ph.D. London, 1873-76, i. 148.

93 Dennys' *Folk-Lore of China*, p. 117.

94 *The Middle Kingdom*, by S. Wells Williams, LL.D. New York, 1883, ii. 74.

95 *The Disowned*, by the Right Hon. Lord Lytton, chap. lxii.

96 *Fiji and the Fijians*, by Thomas Williams. London, 1858, i. 205.

97 *Primitive Culture*, i. 321.

98 *On the Aborigines of Southern Australia*, by W. E. Stanbridge, of Wombat, Victoria. Transactions of Ethnolog. Society of London, 1861, p. 301.

99 *A Discovery of a New World*, by John Wilkins, Bishop of Chester. London, 1684, p. 77.

100 *A Voyage to the Pacific Ocean*, by Capt. James Cook, F.R.S., and Capt. James King, LL.D., F.R.S. London, 1784, ii. 167.

101 *Polynesian Researches during a Residence of nearly Eight Years in the Society and Sandwich Islands*, by William Ellis. London, 1833, iii. 171.

102 *Prehistoric Times*, by Sir John Lubbock, Bart., M.P., D.C.L. London, 1878, p. 440.

103 *Primitive Culture*, i. 318.

104 *See* Kalisch on *Genesis*. London, 1858, p. 70.

105 *Sermons*, by the Rev. W. Morley Punshon, LL.D. Second Series. London, 1884, p. 376.

106 *Outlines of the History of Religion*, by C. P. Tiele. Trans. by J. E. Carpenter. London, 1877, p. 8.

107 *The Myths of the New World*, by Daniel G. Brinton, A.M., M.D. New York, 1868, p. 131.

108 *The Primitive Inhabitants of Scandinavia.* By Sven Nilsson (Lubbock's edit.). London, 1868, p. 206.

109 *Lectures on the Science of Language.* London, 1880, i. 6.

110 *Teutonic Mythology,* iii. 704.

111 *The Manners and Customs of the Ancient Egyptians.* London, 1878, iii. 39.

112 *Ibid.*, iii. 165.

113 *The Mythology of the Aryan Nations.* London, 1882, note to p. 372.

114 *Russian Folk-Lore,* by W. R. S. Ralston, M.A. London, 1873, p. 176.

115 Tylor's *Primitive Culture,* i. 260.

116 *A System of Biblical Psychology,* by Franz Delitzsch, D.D., translated by the Rev. R. E. Wallis, Ph.D. Edinburgh, 1875, p. 124.

117 *The Book of Isaiah* liv. 4-6, and lxii. 4.

118 *English Grammar, Historical and Analytical,* by Joseph Gostwick. London, 1878, pp. 67-72.

119 *Hibbert Lectures* for 1878, p. 190.

120 Bayle's *Dictionary,* i. 113.

121 Vide Tylor's *Anthropology.* London, 1881, p. 149.

122 *Language and Languages,* by the Rev. Frederic W. Farrar, D.D., F.R.S. London, 1878, p. 181.

123 *Ibid.*, p. 182. Coleridge also was in error on this question. See his *Table Talk,* under date May 7th, 1830.

124 *Hebrew and Christian Records,* by the Rev. Dr. Giles. London, 1877, i. 366.

125 *Biblical Psychology*, p. 79.

126 *Antitheism*, by R. H. Sandys, M.A. London, 1883, p. 32.

127 *The Origin and Development of Religious Belief.* London, 1878, i. 187.

128 *The Works of Ralph Waldo Emerson*. London, 1882, i. 274.

129 *Jesus Christ: His Times, Life, and Work*, by E. de Pressensé. London, 1866, p. 38.

130 *Sketches of the History of Man*, by the Hon. Henry Home of Kames. Edinburgh, 1813, iii. 364.

131 *Ancient Faiths Embodied in Ancient names*, by Thomas Inman. London, 1872, ii. 325.

132 *Mythology among the Hebrews*, by Ignaz Goldziher, Ph.D. London, 877, p. 76.

133 *Primitive Culture*, ii. 271.

134 *Nineveh and its Remains*, by Austen Henry Layard, M.P. London, ii. 446.

135 Inman's *Ancient Faiths*, i. 93.

136 *The Unicorn: a Mythological Investigation*, by Robert Brown, F.S.A. London, 1881, p. 34.

137 *The Five Great Monarchies of the Ancient Eastern World*, by George Rawlinson, M.A. London, 1871, i. 56.

138 *Ibid.*, vol. i. p. 123.

139 *Ibid.*, vol. i. note to p. 124.

140 Brown's *Unicorn*, p. 34.

141 *Mythology among the Hebrews*, p. 158,

142 *Ibid.*, 159.

143 *Ibid.*, 160.

144 *Jewish History and Politics*, by Sir Edward Strachey, Bart. London, 1874, p. 256.

145 *Phoenicia*, by John Kenrick, M.A. London, 1855, p. 301.

146 *Dictionary of the Bible*, edited by William Smith, LL.D. Art. ASHTORETH.

147 *Dictionary of the Scottish Language*, iii. 299.

148 On *Isaiah*. London, 1824, ii. 374.

149 *The Antiquities of Israel*, by Heinrich Ewald (trans. by Solly). London, 1876, p. 341.

150 *The Bampton Lectures for 1876*, by William Alexander, D.D., D.C.L. London, 1878, p. 378.

151 *Rivers of Life, showing the Evolution of Faiths*, by Major-General J. G. R. Forlong. London, 1883, ii. 62.

152 *Outlines of the History of Religion*, by C. p. Tiele, p. 63.

153 *Dictionary of Phrase and Fable*, p. 194.

154 *The Philosophy of History*, by Frederick von Schlegel, translated by J. B. Robertson. London, 1846, p. 325.

155 *El-Koran; or, The Koran*, translated from the Arabic by J. M. Rodwell, M.A. London, 1876, p. 199.

156 Tylor's *Primitive Culture*, ii. 274.

157 *Decline and Fall of the Roman Empire*. London, 1862, p. 76.

158 *The Zend-Avesta*, translated by James Darmesteter. Oxford, 1883, Part ii., p. 90.

159 *The Philosophy of History*, by G. W. F. Hegel, translated by J. Sibree, M.A. London, 1861, p. 186.

160a *The Highlands of Central India*, by Captain J. Forsyth. London, 1871, p. 146.

160b *Travels from St. Petersburg in Russia to Diverse Parts of Asia*, by John Bell of Antermony. Glasgow, 1763, i. 230.

161 *The Early Races of Scotland*, by Forbes Leslie. Edinburgh, 1866, i. 138.

162 Kames' *History of Man*, iii. 299.

163 *The Origin of Civilization and the Primitive Condition of Man*, by Sir John Lubbock, Bart., M.P., F.R.S., D.C.L., LL.D. London, 1882, p. 315.

164 *History Of Man*, iii. 366.

165 *The Religions of China*, by James Legge. London, 1880, p. 12.

166 *Ibid.*, pp. 44-46.

167 *Religion in China*, by Joseph Edkins, D.D. London, 1878, p. 60.

168 *A Translation of the Confucian Yih King*, by the Rev. Canon McClatchie, M.A. Shanghai, 1876, p. 386.

169 *Ibid.*, p. 388.

170 *Ibid.*, p. 449.

171 *The Religions of China*, p. 170.

172 *Religion in China*, p. 105.

173 *Handbook for the Student of Chinese Buddhism*, by Rev. E. J. Eitel, London. 1870, p. 107.

174 *Hulsean Lectures for 1870*, p. 203.

175 *Hibbert Lectures on Indian Buddhism*, by T. W. Rhys Davids. London, 1881, p. 231.

176 *A View of China for Philological Purposes*, by the Rev. R. Morrison. Macao, 1817, p. 107.

177 Dennys' *Folk-Lore of China*, p. 28.

178 *Travels in Tartary, Thibet, and China during the years 1844-46*, by M. Huc. Translated by W. Hazlitt. London, i. 61.

179 *Social Life of the Chinese*, by Rev. Justus Doolittle. New York, 1867, ii. 65.

180 *China: Its State and Prospects*, by W. H. Medhurst. London, 1838, p. 217.

181 *Ibid.*, p. 188.

182 *Researches into the Physical History of Mankind*, by James Cowles Prichard, M.D., F.R.S. London, 1844, iv. 496-7.

183 Tylor's *Anthropology*, p. 21.

184 *The Historical Library of Diodorus the Sicilian*, Made English by G. Booth. London, 1700, p. 21.

185 *History of Ancient Egypt*, by George Rawlinson, M.A. London, 1881, i. 369.

186 *Records of the Past*, Edited by S. Birch, LL.D., D.C.L., etc. London, vi. iii.

187 *Hibbert Lectures for 1879*, p. 116.

188 *Ibid.*, p. 155.

189 *Ancient Egypt*, i. 373.

190 *Records of the Past*, iv. 53.

191 *Egypt's Place in Universal History*, by Christian C. J. Bunsen, D.Ph., and D.C.L. Translated by C. H. Cottrell, M.A. London, 1848, i. 395.

192 *Hibbert Lectures*, p. 237.

193 *On the Relations between Pasht, the Moon, and the Cat, in Egypt.* Transactions of the Society of Biblical Archaeology, 1878, vol. vi. 3 16.

194 *Travels to Discover the Source of the Nile, in the years* 1768-73, by James Bruce, F.R.S. Edinburgh, 1813, vi. 343.

195 *Ibid.*, iv. 36.

196 *A Voyage to Congo*, by Father Jerom Merolla da Sorrento. Pinkerton's *Voyages and Travels*. London, 1814, vol. xvi. 273.

197 *Journal of the Anthropological Institute*, May, 1884.

198 *Travels in the Interior Districts of Africa*, by Mungo Park, Surgeon. London, 1779, vol. i. 271.

199 *Ibid.*, i. 322.

200 *Missionary Travels and Researches in South Africa*, by David Livingstone, LL.D., D.C.L., etc. London, 1857, p. 235.

201 *The Present State of the Cape of Good Hope*, by Peter Kolben, A.M. London, 1731, i. 96.

202 *The Poetical Works of Lord Byron*. London, 1876 (*Don Juan*, Canto iii.), p. 636.

203 *The Iliad of Homer.* Translated by J. G. Cordery. London, 1871, ii. 183.

204 *A History of Greece*, by George Grote, F.R.S. London, 1872, i. 317.

205 *Vide Pausan.*, L. x. c. 32, p. 880. Edit. Kuhnii, fol. Lips, 1696.

206 *History of Greece*, i. 317.

207 *The Iliad of Homer*, by Edward Earl of Derby. London, 1867, i. 190.

208 See *Roman Antiquities*, by Alexander Adam, LL.D. London, 1825, pp. 251-60.

209 *Carmen Saeculare*, 35.

210 *Metam.*, lib. xi. 657.

211 *The Divine Legation of Moses Demonstrated*, by William Warburton, D. D. London, 1837, i. 316.

212 Jamieson's *Scottish Dictionary*, iii. 299.

213 *Teutonic Mythology*, ii. 704.

214 *Chaldaean Magic: Its Origin and Development*, by François Lenormant. London, p. 249.

215 *Flammarion's Astronomical Myths*, p. 35.

216 Leslie's *Early Races of Scotland*, i. 113.

217 *Ibid.*, i. 134.

218 *Remaines of Gentilisme and Judaisme*, by John Aubrey, 1686-7. Edited by James Britten, F.L.S. London, 1881, p. 83.

219 *Britannia*, by William Camden, translated by Edmund Gibson, D.D. London, 1772, ii. 380.

220 *A General History of Ireland from the Earliest Accounts*, by Mr. O'Halloran. London, 1778, i. 47.

221 *Ibid.*, i. 113.

222 *Ibid.*, i. 221.

223 *The Towers and Temples of Ancient Ireland*, by Marcus Keane, M.R.I.A. Dublin, 1867, p. 59.

224 *The Keys of the Creeds*. London, 1875, p. 148.

225 A. S., in *Notes and Queries* for Nov. 19, 1881, p. 407.

226 *History of the Missions of the United Brethren among the Indians in North America*, by George Henry Loskiel. London, 1794, Part i. p. 40.

227 *Illustrations of the Manners, Customs, and Condition of the North American Indians*, by George Catlin. London, 1876, ii. 242.

228 *Scenes and Studies of Savage Life*, by Gilbert Malcolm Sproat. London, 1868, p. 206.

229 Brown's *Races of Mankind*, p. 142.

230 Lubbock's *Origin of Civilization*, p. 315.

231 See *Mexico To-day*, by Thomas Unett Brocklehurst. London, 1883, p. 175.

232 Bancroft's *Races of the Pacific*, i. 587.

233 *Ibid.*, iii. 112.

234 *Ibid.*, iii. 187.

235 *Hibbert Lectures for 1884*, p. 45.

236 *American Antiquities and Researches into the Origin anti History of the Red Race*, by Alexander W. Bradford. New York, 1843, p. 353.

237 *Travels in Brazil in the Years* 1817-20, by Dr. Joh. Bapt. von Spix and Dr. C. F. Phil. von Martius. London, 1824, ii. 243.

238 *An Account of the Abipones, an Equestrian People of Paraguay*, from the Latin of Martin Dobrizhoffer. London, 1822, ii. 65.

239 *The Royal Commentaries of Peru*, by the Inca Garcilasso de la Vega. Translated by Sir Paul Rycaut, Knt. London, 1688, folio, p. 455.

240 *Narratives of the Rites and Laws of the Yncas*. Translated from the Spanish MS. of Christoval de Molina, by Clements R. Markham, C.B., F.R.S. London, 1873, p. 37.

241 *History of the Conquest of Peru*, by William H. Prescott. London, 1878, p. 47.

242 *Jottings during the Cruise of H.M.S. Curaçoa among the South Sea Islands* in 1865, by Julius L. Brenchley, M.A., F.R.G.S. London, 1873, p. 320.

243 *Polynesian Mythology*, by Sir George Grey, late Governor in Chief of New Zealand. London, 1855, *Pref.* xiii.

244 Kenrick's *Phoenicia*, p. 303.

245 *Workes* of John Baptista Van Helmont. London, 1644, p. 142.

246 Goldziher's *Hebrew Mythology*, Note to p. 206.

247 *Ibid.*, p. 206.

248 Dr. Smith's *Bible Dictionary*, Article *Meni*, by William A. Wright, M.A., ii. 323.

249 Goldziher's *Hebrew Mythology*, p. 160.

250 Gubernatis' *Zoological Mythology*, i. 18.

251 *Ibid.*, ii. 375.

252 Mayers' *Chinese Reader's Manual*, p. 288.

253 *Japanese Fairy World*. Stories from the Wonder Lore of Japan, by William Elliot Griffis. Schenectady, N. Y., 1880, p. 299.

254 Brown's *Unicorn*, p. 69.

255 Wilkinson's *Ancient Egyptians*, iii. 375.

256 *Teutonic Mythology*, ii. Note to p. 719.

257 Brinton's *Myths of the New World*, p. 130.

258 Schoolcraft's *Indian Tribes*, iii. 485.

259 *Myths of the New World*, p. 133.

260 *Ibid.*, p. 134.

261 *Origin of Civilization*, p. 315.

262 *Myths of the New World*, pp. 135-7.

263 *Ibid.*, p. 131.

264 Tylor's *Primitive Culture*, i. 318.

265 Chambers's *Etymological Dictionary* (Findlater).

266 *Dictionary of Phrase and Fable*, p. 865.

267 *Ecclesiastical Polity*. London, 1617, p. 191.

268 *The Natural History of Infidelity and Superstition*, by J. E. Riddle, M.A. Oxford, 1852, p. 155.

269 *The Anatomy of Melancholy*. London, 1836, p. 669.

270 *The Descent of Man*, by Charles Darwin, M.A., F.R.S., etc. London, 1877, p. 121.

271 *Essays. Of Superstition.*

272 *Lectures on the Philosophy of the Human Mind*. Edinburgh, 1828, p. 673

273 *Voltaire*, by John Morley. London, 1878, p. 156. See also Parton's *Life of Voltaire*.

274 Gubernatis' *Zoological Mythology*, i. 56.

275 *Vide* Inman's *Ancient Faiths*, ii. 260, 326.

276 Mosheim's *Ecclesiastical History*. London, 1847, i. 116.

277 *History of Brazil*, by Robert Southey. London, 1810, p. 635.

278 *The Dictionary, Historical and Critical*. London, 1734, iv. 672.

279 *Primitive Culture*, i. 262.

280 Leslie's *Early Races of Scotland*, ii. 496.

281 *History of Brazil*, i. 193.

282 *Icelandic Legends*. Collected by Jón Arnason (Powell and Magnússon). London, 1866, p. 663.

283 *On the Truths contained in Popular Superstitions*, by Herbert Mayo, M.D. Edinburgh and London, 1851, p. 135.

284 *A Literal Translation of Aristophanes: The Clouds*, by a First-Class Man of Balliol College. Oxford, 1883, p. 31.

285 See *Curiosities of Indo-European Tradition and Folk-Lore*, by Walter H. Kelly. London, 1863, p. 226.

286 *Teutonic Mythology*, ii. 706.

287 *Astronomical Myths*, p. 331

288 *Medea: a Tragedie*. Written in Latin by Lucius Anneus Seneca. London, 1648, p. 105.

289 *The Childhood of the World*, by Edward Clodd, F.R.A.S. London, 1875, p. 65.

290 *The Chinese Empire*, by M. Hue. London, 1855, ii. 376.

291 *The Connection of the Physical Sciences*. London, 1877, p. 104.

292 Grimm's *Teutonic Mythology*, ii. 707.

293 *Appendix on the Astronomy of the Ancient Chinese*, by the Rev. John Chalmers, A.M. Legge's Chinese Classics. Vol. iii. Part i. Hong-Kong, 1861, p. 101.

294 *The Middle Kingdom*, i. 818.

295 *Ibid.*, ii. 73.

296 *Social Life of the Chinese*, by the Rev. Justus Doolittle, of Fuhchau. New York, 1867, i. 308.

297 *Chinese Sketches*, by Herbert A. Giles. London, 1876, p. 99.

298 *Gems of Chinese Literature*, by Herbert A. Giles. Shanghai, 1884 p. 102.

299 *An Account of Cochin China*. Written in Italian by the R. E. Christopher Borri, a Milanese, of the Society of Jesus. Pinkerton's Travels, ix. 816.

300 *A Voyage to and from the Island of Borneo in the East Indies*, by Captain Daniel Beeckman. London, 1878, p. 107.

301 *History of the Indian Archipelago*, by John Crawfurd, F.R.S. Edinburgh, 1820, i. 305.

302 *Sketches of the History of Man*, iii. 300.

303 *Thucydides*. Translated by B. Jowett, M.A. Oxford, 1881, i. 521.

304 *The Stratagems of Jerusalem*, by Lodowick Lloyd, Esq., One of her Majestie's Serjeants at arms. London, 1602, p. 286.

305 Quoted in *Notes and Queries*, 16th of April, 1881, by William E. A. Axon.

306 *Northern Antiquities*, by Paul Henri Mallett. London, 1790, i. 39.

307 *Teutonic Mythology*, i. 245.

308 *Ibid.*, ii. 705.

309 *Advice to a Son*. Oxford, 1658, p. 105

310 Grimm's *Teutonic Mythology*, ii. 714.

311 Schoolcraft's *Indian Tribes*, v. 2 16.

312 Brinton's *Myths*, p. 137.

313 Bradford's *American Antiquities*, p. 332.

314 *Ibid.*, p. 333.

315 *The Antiquities of Mexico*, by Augustine Aglio. London, 1830, folio vi. 144.

316 Bancroft's *Native Races*, iii. 111.

317 Brinton's *Myths*, p. 131.

318 *Polynesian Researches*, i. 331.

319 Mariner's *Natives of the Tonga Islands*, ii. 127.

320 *Discoveries in the Ruins of Nineveh and Babylon*. London, 1853, p. 552.

321 Tylor's *Primitive Culture*, ii. 272.

322 *Ibid.*, ii. 272.

323 *Description of the Western Islands of Scotland*, by Martin Martin. London, 1716, p. 41.

324 *The Philosophie*, p. 696.

325 *A Voyage to St. Kilda, the remotest of all the Hybrides*, by M. Martin, Gent. Printed in the year 1698. Miscellanea Scottica. Glasgow, 1818, p. 34.

326 *The Zend-Avesta*. Oxford, 1883, ii. 90.

327 *Five Hundred pointes of good Husbandrie*, by Thomas Tusser. London, 1580, p. 37.

328 Flammarion's *Marvels of the Heavens*, p. 244.

329 *The Philosophie*, 1603, p. 697.

330 *English Folk-Lore*, by the Rev. T. F. Thiselton Dyer, M.A., Oxon. London, 1880, p. 42.

331 *Notes on the Folk-Lore of the Northern Counties of England and the Borders*, by William Henderson. London, 1866, p. 86.

332 *Knowledge for the Time*, by John Timbs, F.S.A. London, p. 227.

333 *Popular Errors, Explained and Illustrated*, by John Timbs, F.S.A. London, 1857, p. 131.

334 *A Manual of Astrology*, by Raphael. London, 1828, p. 90.

335 Brinton's *Myths*, p. 132.

336 *Endimion: The Man in the Moone*. London, 159 1, Act i. Sc. I.

337 *A defensative against the poyson of supposed Prophecies*, by Henry Howard, Earl of Northampton. London, 1583.

338 *Folk-Lore of China*, p. 118.

339 Tusser's *Good Husbandrie*, p. 13.

340 *Ibid.*, p. 13.

341 *Folk-Lore of the Northern Counties of England*, p. 41

342 *David Copperfield*. The "Charles Dickens" edition, p. 270.

343 See *An Historical Survey of the Astronomy of the Ancients*, by the Rt. Hon. Sir George C. Lewis, Bart. London, 1862, p. 312.

344 *Popular Astronomy*, by Simon Newcomb, LL.D. New York, 1882, p. 325.

345 *Primitive Culture*, i. 118.

346 Dennys's *Folk-Lore of China*, p. 32.

347 *Folk-Lore; or, Manners and Customs of the North of England*, by M.A.D. Novo-Castro-sup. Tynan, 1850-51, p. 11.

348 Dyer's *Folk-Lore*, p. 42.

349 *Ibid.*, p. 41.

350 *Time's Telescope* for 1814. London, p. 368.

351 Dennys's *Folk-Lore of China*, p. 118.

352 *The Book of Days: a Miscellany of Popular Antiquities*. Edited by R. Chambers. London and Edinburgh, ii. 203.

353 *The Life and Correspondence of Robert Southey*. Edited by his son. London, 1850, v. 341.

354 *Adam Bede*, chap. xviii.

355 *Scottish Ballads and Songs*. Edited by James Maidment. Edinburgh, 1868, i. 41.

356 *Etymological Dictionary of the Scottish Language*. Paisley, 1880, iii. 299.

357 Dyer's *Folk-Lore*, p. 38.

358 *Notes and Queries* for May 16th, 1874, p. 384.

359 *Ibid.* for August 1st, 1874, p. 84.

360 *Amazulu*, by Thomas B. Jenkinson, B.A., late Canon of Maritzburg. London, 1882, p. 61.

361 *Legends of Iceland*. Collected by Jón Arnason. Second series. London, 1866, p. 635.

362 *Astrology, as it is, not as it has been represented*, by a Cavalry Officer. London, 1856, p. 37.

363 *A Manual of Astrology*, by Raphael. London, 1828, p. 89.

364 *The Dignity and Advancement of Learning*. London (Bohn), 1853, p. 129.

365 *Works*. London, 1740, iii. 187.

366 Dyer's *Folk-Lore*, p. 41.

367 *Scottish Dictionary*, iii. 300.

368 Tylor's *Primitive Culture*, i. 117.

369 Vide Potter's *Antiquities of Greece*, ii. 262.

370 *Recreations in Astronomy*, by the Rev. Lewis Tomlinson, M.A. London, 1858, p. 251.

371 Flammarion's *Marvels of the Heavens*, p. 243.

372 *Genesis, with a Talmudic Commentary*, by Paul Isaac Hershon. London, 1883, p. 50.

373 *Notes on the Miracles*, p. 363.

374 *The Gospel of S. Matthew illustrated from Ancient and Modern Authors*, by the Rev. James Ford, M.A. London, 1859, p. 310.

375 See *Light: Its Influence on Life and Health*, by Forbes Winslow, M.D., D.C.L. London, 1867, p. 94. Also, *The History of Astronomy*, by George Costard, M.A. London, 1767, p. 275.

376 *The Science and Practice of Medicine*, by William Aitken, M.D. London, 1864, ii. 353.

377 *Myths of the New World*, p. 132.

378 *Ibid.*, p. 134.

379 *Ibid.*, p. 135.

380 *The Darker Superstitions of Scotland illustrated from history and practice*, by John Graham Dalyell. Edinburgh, 1834, p. 286.

381 *The Early Races of Scotland*, i. 136.

382 *The Statistical Account of Scotland*, by Sir John Sinclair, Bart. Edinburgh, 1791, i. 47.

383 *Works*. London, 1740, iii. 187.

384 Dyer's *Folk-Lore*, p. 47.

385 *Some West Sussex Superstitions Lingering in* 1868. Collected by Charlotte Latham, at Fittleworth. *The Folk-Lore Record* for 1878, p. 45.

386 Dyer's *Folk-Lore*, p. 48.

387 Inman's *Ancient Faiths*, ii. 327.

388 *Fairy Tales: their origin and meaning*, by John Thackray Bunce. London, 1878, p. 131.

389 Martin's *Western Islands of Scotland*, 1716, p. 42.

390 *Letters from the East*, by John Carne, Esq. London, 1826, p. 77.

391 Grimm's *Teutonic Mythology*, ii. 715.

392 Timbs's *Knowledge for the Time*, p. 227.

393 *Dissertation upon Superstitions in Natural Things*, by Samuel Werenfels, Basil, Switzerland. London, 1748, p. 6.

394 Vide Grimm's *Teutonic Mythology*, ii. 714-716.

395 *Defensative*, 1583.

396a *A Talmudic Miscellany*. Compiled and translated by Paul Isaac Hershon. London, 1880, p. 342.

396b *Caesar's Commentaries*. London (Bohn), 1863, Book i. Chap. 50.

397 *Popular Rhymes*, p. 217.

398 *The Year Book of Daily Recreation and Information*, by William Hone. London, 1838, p. 254.

399 Dyer's *Folk-Lore*, p. 43.

400 *Gentilisme*, p. 37.

401 Dyer's *Folk-Lore*, p. 44.

402 *Extraordinary Popular Delusions*. London, i. 260.

403 Dyer's *Folk-Lore*, p. 38.

404 Henderson's *Folk-Lore*, p. 86.

405 *Popular Romances of the West of England*. Collected by Robert Hunt, F.R.S. London, 1881, p. 429.

406 *West Sussex Superstitions*, p. 10.

407 C. W. J. in Chambers's *Book of Days*, ii. 202.

408 *Early Races of Scotland*, i. 136.

409 *Scottish Dictionary*, iii. 300.

410 Forlong's *Rivers of Life*, ii. 63.

411 *Secret Memoirs of the late Mr. Duncan Campbel*. Written by Himself. London, 1732, p. 62.

412 *Folk-Lore*, 1851, p. 8.

413 *Popular Rhymes.*

414 Jamieson's *Scottish Dictionary*, iii. 300.

415 *Familiar Illustrations of Scottish Character*, by the Rev. Charles Rogers, LL. D. London, 1865, p. 172.

416 *Statistical Account of Scotland*, xii. 457.

417 *Early Races of Scotland*, ii. Note to p. 406.

418 Dalyell's *Darker Superstitions of Scotland*, p. 285.

419 *Romeo and Juliet*, Act ii. Sc. 2.

420 *Light: Its Influence on Life and Health*, p. 101.

421 *Religion as Affected by Modern Materialism*, by James Martineau, LL.D. London, 1874, pp. 7, 11.

422 *The Relations between Religion and Science*. Bampton Lectures for 1884, p. 245.

423 *Address delivered before the British Association assembled at Belfast*, by John Tyndall, F.R.S. London, 1874, p. 61.

424 *Celestial Objects for Common Telescopes*, by the Rev. T. W. Webb, M.A., F.R.A.S. London, 1873, p. 58.

425 *The Heavens*, by Amédée Guillemin. London, 1876, p. 144.

426 *McFingal*, by John Trumbull. Hartford, U.S.A., 1782 Canto i. line 69.

427 *Stargazing*, by J. Norman Lockyer, F.R.S. London, 1878, p. 476.

428 *The System of the World*, by M. le Marquis de La Place. Dublin, 1830, i. 42.

429 *The Solar System*, by J. Russell Hind. London, 1852, p. 48.

430 *History of Physical Astronomy*, by Robert Grant, F.R.A.S. London, 1852, p. 230.

431 *Cosmos*, by Alexander von Humboldt (Sabine's Edition). London, 1852, iii. 357.

432 *Handbook of Astronomy*, by Dionysinus Lardner, D.C.L. London, 1853, pp. 194, 197.

433 *The Planetary Worlds*, by James Breen. London, 1854, p. 123.

434 *Of the Plurality of Worlds. An Essay.* Fourth Edition. London, 1855, p. 289.

435 *Vestiges of the Natural History of Creation.* Eleventh Edition. London, 1860, pp. 21, 22.

436 *The Treasury of Science*, by Friedrich Schoedler, Ph.D. London, 1865, p. 167.

437 *Spectrum Analysis*, by Dr. H. Schellen. London, 1872, p. 481.

438 *The Moon*, by James Nasmyth, C.E., and James Carpenter, F.R.A.S. London, 1874, p. 157.

439 *Astronomy*, by J. Rambosson. Translated by C. B. Pitman. London, 1875, p. 191.

440 *The Three Heavens*, by the Rev. Josiah Crampton, M.A. London, 1879, p. 328.

441 *Scientific and Literary Treasury*, by Samuel Maunder. London, 1880, p. 470.

442 *The Mathematical and Philosophical Works of John Wilkins.* London, 1708.

443 *A Plurality of Worlds*, by Bernard le Bovier de Fontenelle. London, 1695, p. 35.

444 *More Worlds than One*, by Sir David Brewster, M.A., D.C.L. London, 1874, pp. 120, 121.

445 *An Introduction to Astronomy*, by John Bonnycastle. London, 1822, p. 367.

446 *Elements of Astronomy*, by John Brinkley, D. D., F.R.S. Dublin, 1819, p. 113.

447 *Celestial Scenery*, by Thomas Dick, LL.D. London, 1838, p. 350.

448 *The Moon*, by Edmund Neison, F.R.A.S. London, 1876, pp. 17, 129.

449 *The Art of Scientific Discovery*, by G. Gore, LL. D., F.R.S. London, 1878, p. 587.

450 *The Moon, her Motions, Aspect, Scenery, and Physical, Condition*, by Richard A. Proctor. London, 1878, p. 300.

451 *Other Worlds than Ours*. London, 1878, p. 167.

452 *An Historical Account of Astronomy*, by John Narrien, F.R.A.S. London, 1833, p. 448. See also Schroeter's, Observations on the Atmosphere of the Moon. Philosophical Trans. for 1792, p. 337.

453 *Plutarch's Morals*. Translated by P. Holland. London, 1603, pp. 825, 1178.

454 *Cosmotheoros*, by Christian Huyghens van Zuylichem. Glasgow, 1757, pp. 177, 178.

455 *The Wisdom of God in the Creation*, by John Ray, F.R.S. London, 1727, p. 66.

456 *On the Earths in our Solar System*, by Emanuel Swedenborg. London, 1840, p. 59.

457 *Philosophical Transactions of the Royal Society for* 1795, p. 66.

458 *Theology*, by Timothy Dwight, LL.D. London, 1836, p. 91.

459 *Astronomy and General Physics*, by William Whewell, M.A. London, 1836, p. 269.

460 *Astronomical Discourses*, by Thomas Chalmers, D. D., LL.D. Edinburgh, 1871, p. 23.

461 *A Defence of Poetry*, in ESSAYS, etc., by Percy Bysshe Shelley. London, 1852, i. 48.

462 *Physical Theory of Another Life*. London, 1836, p. 200.

463 *The Plurality of Worlds, the Positive Argument from Scripture*, etc. London (Bagster), 1855, p. 146.

464 *Elements of Physics*, by Neil Arnott, M.D., F.R.S. London, 1865, part ii. p. 684.

465 *Historical Account of Astronomy*, p. 520.

Printed in Great Britain
by Amazon